World Nitrogen Survey

Industry and Finance Series

Volume 20

This series is produced by the Industry Department of the World Bank to disseminate ongoing work and stimulate further discussion. The series will include studies of individual sectors in industry, aspects of world industry, industrial strategy and policy, and industrial finance and financial development. Already published are the following:

*Volume 1. Structural Changes in World Industry: A Quantitative Analysis of Recent Developments
*Volume 2. Energy Efficiency and Fuel Substitution in the Cement Industry with Emphasis on Developing Countries
*Volume 3. Industrial Restructuring: Issues and Experiences in Selected Developed Economies
*Volume 4. Energy Efficiency in the Steel Industry with Emphasis on Developing Countries
*Volume 5. World Sulphur Survey
*Volume 6. Industrialization in Sub-Saharan Africa: Strategies and Performance
*Volume 7. Small Enterprise Development: Economic Issues from African Experience
*Volume 8. World Refinery Industry: Need for Restructuring
*Volume 9. Guidelines for Calculating Financial and Economic Rates of Return for DFC Projects (also in French and Spanish)
Volume 10. A Framework for Export Policy and Administration: Lessons from the East Asian Experience (also in Spanish)
Volume 11. Fertilizer Producer Pricing in Developing Countries: Issues and Approaches
Volume 12. Iron Ore: Global Prospects for the Industry, 1985-95
Volume 13. Tax Policy and Tax Reform in Semi-Industrial Countries
Volume 14. Interest Rate Policies in Selected Developing Countries, 1970-82
Volume 15. Mobilizing Small Scale Savings: Approaches, Costs, and Benefits
Volume 16. World Bank Lending to Small Enterprises: A Review
Volume 17. Public Industrial Enterprises: Determinants of Performance
Volume 18. High Interest Rates, Spreads, and the Costs of Intermediation: Two Studies
*Volume 19. Credit Guarantee Schemes for Small and Medium Enterprises

* Published as World Bank Technical Papers.

WORLD BANK TECHNICAL PAPER NUMBER 59

INDUSTRY AND FINANCE SERIES

World Nitrogen Survey

William F. Sheldrick

The World Bank
Washington, D.C.

The International Bank for Reconstruction
and Development/THE WORLD BANK
1818 H Street, N.W.
Washington, D.C. 20433, U.S.A.

Manufactured in the United States of America
First printing April 1987

Technical Papers are not formal publications of the World Bank, and are circulated to encourage discussion and comment and to communicate the results of the Bank's work quickly to the development community; citation and the use of these papers should take account of their provisional character. The findings, interpretations, and conclusions expressed in this paper are entirely those of the author(s) and should not be attributed in any manner to the World Bank, to its affiliated organizations, or to members of its Board of Executive Directors or the countries they represent. Any maps that accompany the text have been prepared solely for the convenience of readers; the designations and presentation of material in them do not imply the expression of any opinion whatsoever on the part of the World Bank, its affiliates, or its Board or member countries concerning the legal status of any country, territory, city, or area or of the authorities thereof or concerning the delimitation of its boundaries or its national affiliation.

Because of the informality and to present the results of research with the least possible delay, the typescript has not been prepared in accordance with the procedures appropriate to formal printed texts, and the World Bank accepts no responsibility for errors. The publication is supplied at a token charge to defray part of the cost of manufacture and distribution.

The most recent World Bank publications are described in the catalog *New Publications,* a new edition of which is issued in the spring and fall of each year. The complete backlist of publications is shown in the annual *Index of Publications,* which contains an alphabetical title list and indexes of subjects, authors, and countries and regions; it is of value principally to libraries and institutional purchasers. The continuing research program is described in *The World Bank Research Program: Abstracts of Current Studies,* which is issued annually. The latest edition of each of these is available free of charge from the Publications Sales Unit, Department F, The World Bank, 1818 H Street, N.W., Washington, D.C. 20433, U.S.A., or from Publications, The World Bank, 66, avenue d'Iéna, 75116 Paris, France.

William F. Sheldrick is a fertilizer adviser in the Industry Department of the World Bank.

Library of Congress Cataloging-in-Publication Data

Sheldrick, William F.
World nitrogen survey.

(Industry and finance series ; v. 20) (World Bank technical paper, ISSN 0253-7494 ; no. 59)
1. Nitrogen industries. 2. Fertilizer industry. 3. Nitrogen. I. Title. II. Series. III. Series: World Bank technical paper ; no. 59.
HD9660.N7S55 1987 338.4'7668624 87-6203
ISBN 0-8213-0895-5

ABSTRACT

Nitrogen is an important raw material for the chemical and fertilizer industries. Almost all nitrogen products are derived from ammonia, and about 85% go to fertilizer use. This report examines the world nitrogen industry and its outlook through 1995/96.

The nitrogen fertilizer industry grew very rapidly during the 1960s, when fertilizer consumption increased rapidly as part of the "Green Revolution" to meet increasing world food needs. There was also a major change in the structure of the industry, as both consumption and production increased more rapidly in the Developing Market Economies and Centrally Planned Economies than in the Developed Market Economies. In the first half of the 1980s, however, nitrogen fertilizer growth rates have declined due to a depressed world agricultural situation, and this has resulted in a surplus supply of nitrogen fertilizers with very low prices. Most of this surplus is in the Eastern European countries, who are now the biggest net exporters of nitrogen fertilizer products.

During the next decade, world total nitrogen consumption is projected to grow at an average rate of 2.4% per year and world fertilizer nitrogen at 2.6% per year. In the Developed Market Economies the growth is projected to be about 0.8%, in the Developing Market Economies about 5.2%, and in the Centrally Planned Economies about 2.3% per year.

In a period of oversupply and with a depressed agricultural market, nitrogen fertilizer prices are expected to remain low. After 1988/89, however, the surplus balance will diminish, and prices are forecast to rise sharply and peak in the early 1990s. But they will probably decline again toward the mid-1990s after new capacity comes on stream.

A detailed examination has been made of the total costs of producing urea and ammonia and how this is likely to be affected by future energy prices, plant location, and utilization rates. After 1990 it is estimated that at least 10 new nitrogen fertilizer plants will be needed each year through 1995 to meet increasing demand and to replace worn-out plants.

ACKNOWLEDGEMENT

The author acknowledges the contribution of his colleagues in the World Bank/FAO/UNIDO Fertilizer Working Group in providing much of the data base used in the report and to members of the International Nitrogen Industry who reviewed the early draft. The author is grateful to Nayyer Nizam for his assistance in the statistical analysis work and in particular to Mina Salehi for her assistance in processing the Working Group data and for her patience and care during the final preparation of the study.

TABLE OF CONTENTS

ANNEXES

TABLES

Figures

PRINCIPAL ABBREVIATIONS AND ACRONYMS USED

N	-	Nitrogen equivalent in nitrogenous materials
CAN	-	Calcium Ammonium Nitrate
AS	-	Ammonium Sulphate
DAP	-	Diammonium Phosphate
LPG	-	Liquified Petroleum Gas
LNG	-	Liquified Natural Gas
MM	-	Million
BBL	-	Barrel
BCM's	-	Billion Cubic Meters
BTU's	-	British Thermal Units
tpd	-	Tons Per Day (metric unless otherwise stated)
tpy	-	Tons Per Year (metric unless otherwise stated)
fob	-	Free on Board
cif	-	Cost, Insurance and Freight
CPE's	-	Centrally Planned Economies
ME's	-	Market Economies
FAO	-	Food and Agricultural Organization of the United Nations
UNIDO	-	United Nations Industrial Development Organization
ASEAN	-	Association of South-East Asian Nations
TFI	-	The Fertilizer Institute of the USA
FAI	-	Fertilizer Association of India
TVA	-	Tennessee Valley Authority
IRR	-	Internal Rate of Return

Time Reference

Most statistical data presented in this review use the FAO fertilizer year 1 July - 30 June. For countries that report their fertilizer statistics on a calendar year basis, data are shown under the fertilizer year, the first part of which corresponds to the calendar year i.e. 1984 data are under 1984/85.

SUMMARY AND CONCLUSIONS

Nitrogen in its many forms is an important raw material for the fertilizer and chemical industry. About 85% of nitrogen is used to produce nitrogen fertilizers and the remainder goes into a wide range of industrial products such as synthetic fibres, explosives, animal feed-stuffs etc. Almost all nitrogen products are manufactured from ammonia, which is derived from the catalytic fixation of nitrogen from the atmosphere with hydrogen, obtained from a variety of hydrocarbons, but mainly natural gas. In 1984/85 about 88 million tons of ammonia (expressed in terms of the elemental form N) were produced.

The main growth in nitrogen consumption and production took place during the 1960s and 1970s when there was a rapid increase in fertilizer use, particularly in developing countries. The green revolution, with the increasing use of high yielding varieties of cereals, accelerated the growth of nitrogen fertilizers, particularly urea. Between 1960 and 1970, world-wide growth rates for nitrogen consumption averaged 10% per year (py). During this period there was also a major change in the structure of the industry as both consumption and production increased at a much higher rate in the Developing Market Economies and Centrally Planned Economies than in the Developed Market Economies. About 42% of the world nitrogen industry is now in the Centrally Planned Economies, 41% in the Development Market Economies and 17% in the Developing Market Economies.

A summary of country profiles in which the major aspects of nitrogen fertilizer consumption, production, trade and feedstock availability are briefly discussed is given in Annex 20. This section of the report is mainly for reference purposes and the leading data from this are incorporated in the main body of the review.

Regional and world supply demand balances are based on those prepared by the World Bank/FAO/UNIDO/Industry Fertilizer Working Group, June 1986. Country ammonia capacities and supply capabilities have been derived from the Working Group's plant lists. Regional demand is assessed by the Group on a "Delphi" principle but individual contributions are based on a range of methodology including trend projections, econometric modelling, agricultural programs and sales forecasts. The Fertilizer Working Group's results are in good agreement with the forecasts of the Economic and Projections Department of the World Bank using their Integrated Agricultural/Fertilizer Model.

World Nitrogen Supply

Before 1960 almost all ammonia capacity and production were in the Developed Market Economies. The principal producers were the USA, Western Europe and Japan who together accounted for 90% of the 15 million tons of nitrogen produced at that time. Between 1960 - 1975 there was a rapid increase in nitrogen production in all parts of the world, but since that period the most rapid growth has been in the Centrally Planned Economies of USSR and China and in some of the major Developing Market Economies such as India, Brazil and Indonesia. The development of the capacity of major nitrogen producers since 1975 is indicated below.

Major Country Ammonia Capacities (Million tons N per year)

Country	1975/76	1980/81	1985/86 (estimated)	1990/91 (projected)
USSR	11.6	18.2	24.5	24.9
USA	14.2	15.6	13.6	13.7
China	6.0	16.0	16.9	18.3
India	2.6	4.8	7.0	9.8

In the last few years both USSR and China have overtaken the USA in ammonia capacity and production. The USA and several countries in Western Europe have reduced their ammonia production recently in the face of competition from imported materials. World-wide, there was a very large construction program in new ammonia plants during the 1970s but in the face of declining nitrogen fertilizer consumption growth rates, the construction of new plants has declined since 1980 and between 1985-89 new capacity erected will be less than any other five year period since 1965. This may well have an important impact on the nitrogen supply situation after 1989.

Ammonia Plants Commissioned World-Wide within 5 Year Time Periods

Year	1960-64	1965-69	1970-74	1975-79	1980-84	1985-89 a/
No. of New Plants	49	115	106	121	87	55
Million Tons Capacity	4.99	18.21	21.2	27.6	18.4	14.2
Average Plant Size 000 tpy	102	158	200	228	211	259

a/ Projected.

These figures, which exclude the small Chinese plants constructed during this period, also show how the average capacity of ammonia plants has increased from about 100,000 tpy, to nearly 260,000 tpy since 1960. The development of the large ammonia plant with centrifugal compressor and improved energy efficiency has played a major part in reducing ammonia production costs and prices.

World Nitrogen Consumption

World nitrogen consumption has increased about 10 fold over the last three decades. There have been two major recessions in demand, the first took place in 1975 after a major price hike during the so called "Fertilizer Crisis". The second occurred in early 1980s as a result of the global recession and the "Payment in Kind" (P.I.K) program in the USA.

World Nitrogen Consumption 1955-1995 (Million tons N)

	54/55	59/60	64/65	69/70	74/75	79/80	84/85	89/90	94/95
Fertilizer N	6.2	9.2	15.3	28.7	38.6	57.3	70.5	80.4	90.7
Industrial N	1.5	3.9	6.0	7.0	8.1	9.2	10.3	11.1	12.3
Total	7.7	13.1	21.3	35.7	46.7	66.5	80.8	91.5	103.0
Average % Yearly Increase Over 5 Year Period		11.2	10.2	10.9	5.5	7.3	4.0	2.6a/	2.2a/

a/ Forecast.

Between 1984/85 and 1995/96 total world nitrogen consumption is forecast to increase on average at 2.4% py.

Between 1955 and 1975 nitrogen fertilizer growth rate averaged about 10% py but this rate has been falling steadily since and between 1984/85 and 1995/96 is expected to average only about 2.6%. In the Developed Market Economies, mainly the USA and Western Europe, due to a depressed farming situation and also a mature fertilizer market, the growth rate is forecast to be less than 1% py, in the Developing Market Economies about 5% and in the Centrally Planned Economics 2.5%.

Nitrogen Fertilizer Demand (Million tons N)

	1984/85	1985/86	1986/87	1987/88	1988/89	1989/90	1990/91	1995/96
Developed M.E.	23.7	23.1	22.7	23.0	23.4	24.4	24.6	26.0
Developing M.E.	15.8	16.5	17.6	18.6	19.8	21.0	22.2	27.5
Central. Plan. Ec.	30.9	30.5	31.6	32.7	33.8	35.0	36.2	39.9
World	70.5	70.2	71.8	74.4	77.0	80.4	83.0	93.3

Industrial nitrogen demand in 1984/85 totalled 10.3 million tons (13% of total nitrogen use) and is expected to grow at about 2.0% py over the next decade. About 68% of industrial nitrogen demand for ammonia occurs in the Developed Market Economies, 26% in the Centrally Planned Economies and 6% in the

Developing Market Economies. The four major consumers are USA, USSR, Japan and Western Europe. In Japan about 60% of all nitrogen products consumed goes to industrial use.

A comparison of the supply and demand for nitrogen, which has been made by an aggregation of detailed country by country figures, shows that the current world surplus of nitrogen is likely to continue for two or three years and then due to a fall-off in the construction of new capacity the surplus will diminish sharply, resulting in a tightly balanced situation by 1989/90.

World Nitrogen Balances (Million tons of N)

	1985/86	1986/87	1987/88	1988/89	1989/90	1990/91
Developed M.Ec.	-0.20	0.52	0.81	0.60	-0.26	-0.40
Developing M.Ec.	-1.44	-1.41	-1.38	-1.57	-1.53	-1.51
Centrally Planned Ec.	4.97	4.74	4.38	3.71	2.58	1.37
World	3.33	3.84	3.86	2.74	0.79	-0.54

The world fertilizer industry is well-known for its cyclical behavior, as supply/demand balances have fluctuated considerably throughout the years. The balances above however also indicate a changing structure in the nitrogen fertilizer industry. In the Developed Market Economies, production capacity will not increase due to low growth rates of nitrogen demand and relatively high energy costs. The USA and Western Europe in the longer term will increase their dependence on imported ammonia. This trend, however, will slow down in the next two or three years in the USA due to an anticipated fall of domestic nitrogen consumption as the 1985 new farm bill attempts to reduce agricultural surpluses. A weakening US Dollar and lower gas prices will also make domestic production more competitive. On the other hand, the deficit in the Developing Market Economies will remain, as increasing demand keeps pace with new capacity coming on stream in the energy-rich countries of the Near East and in India.

In the Centrally Planned Economies, Eastern Europe, mainly through the USSR and Romania, will have a very large nitrogen surplus on which the world will depend to maintain a balanced situation. China is expected to maintain its dependence on imports of about 2-2.5 million tons of N py.

Due to the relatively long lead time for nitrogen fertilizer projects to be implemented, it will be difficult to realize a significantly higher supply situation than that projected now for 1989/90. It seems highly likely that if total nitrogen demand continues to increase as projected, at about 2.4% py (a relatively pessimistic growth rate compared with previously projected rates), a tight situation will develop in two or three years' time. Unfortunately, very low prevailing prices, and worse prospects for the immediate future are inhibiting further investment in new nitrogen fertilizer plants both in Developed and Developing Market Economies. The USSR, because of its large surplus capacity, will continue to dominate the world export market for many years to come. If for one reason or another the USSR does not maintain this high level of exports, for example, by increasing its own consumption more rapidly, or by failing to bring all its plants to full production as forecast, the tight situation projected for 1989/90 will be further aggravated.

World Nitrogen Fertilizer Trade

Fertilizer trade in nitrogen products and ammonia has grown steadily in all regions in recent years. In the last two decades, trade in nitrogen fertilizers has grown from about 3 million tons to more than 14 million tons.

About 21% of all fertilizer nitrogen production is now exported. In the same period the trade outside the Developed Market Economies has increased from virtually zero to about 40% of all export trade (including ammonia) in 1984/85. This is due mainly to the new export-based plants in the Near East and Eastern Europe (including the USSR). The USA is still the main importing and exporting country for nitrogen materials due to the fact that it imports large quantities of ammonia, most of which are processed and exported as diammonium phosphate. Other large exporters are the USSR, the Netherlands, Romania and Canada. Major importers are China, Germany F.R., India and Iran.

The types of fertilizers traded have also changed considerably in the last two decades. For example, ammonium sulphate and ammonium nitrate, which comprised about 57% of the export trade in the mid-1960s, today only amount to 13%. On the other hand urea, which was near 20% in the mid 1960s, in 1984 was 52% of total nitrogen fertilizer trade. Many of the major exporters of nitrogen fertilizer in the Developed Market Economies are now being displaced by exporting countries elsewhere, who have cheap natural gas. For example, in 1973 Japan had about 30% of nitrogen fertilizer trade and dominated the Far East market. In 1983 it had only 1%, due to the fact that, with expensive feedstocks such as naphtha and imported LPG, it can no longer compete in the export markets. In 1982, the USA became a net importer of nitrogen and Western Europe is also now a net importer.

Although it would seem a natural trend for new export capacity to develop where gas is cheap, this trend may slow down because of the difficulties of financing and establishing export based projects in developing countries where investment costs are high. Also improvements in energy consumption in new ammonia plants, together with prospects of lower energy prices generally, have recently reduced the comparative advantage of export based plants near cheap gas sources, particularly where high investment in infrastructure is needed. Another important development in nitrogen fertilizer trade is the increasing use of countertrade, particularly with Eastern European countries and about 25% of nitrogen fertilizer trade is conducted through various types of barter deals, counterpurchase or buy back arrangements. So long as Eastern Europe retains its dominance in the export market and until there is a major structural change in the balance of trade and debt relationship between the Developed Market Economies and the rest of the world, countertrade seems likely to stay as a significant component of overall nitrogen trade.

Investment and Production Costs for Nitrogen Fertilizers

In order to appreciate more fully the factors which influence fertilizer costs and enable more accurate projections of future fertilizer prices to be made, cost estimates have been calculated for a range of plant locations and feedstocks. Bearing in mind that prices of raw materials and investment costs can vary considerably, "cost envelopes" have been developed for ammonia and urea which enable investment costs, production costs and realization prices to be derived over a wide range of conditions. This information is contained in detail in both graphical and tabular form in the review. The information has been used to assess the competitive situation for a large number of potential ammonia and urea projects based on both greenfield sites and existing fertilizer plant sites. The exercise looks at the inter-relationship of investment cost, plant size, utilization rate and gas pricing. The comparisons show that even with very cheap gas, say less than US$1.0/MMBTU, export-based projects are only likely to compete in the overseas nitrogen fertilizer markets if their investment costs can be limited. For example, based on present price forecasts, in the case of a 1,000 tpd export based ammonia plant, the total capital cost, including infrastructure needs, should not exceed about US$200 million and in the case of a 1,700 tpd export based urea complex, the investment cost should not exceed about US$320 million. Unfortunately, very few plants can be built in developing countries on a new greenfield site for this order of cost.

World Nitrogen Fertilizer Plant Capacities and Investment Needs 1990-1995

In 1990/91 it is projected that world nitrogen supply and demand will be in deficit. World nitrogen fertilizer consumption is expected to increase by about 10.3 million tons between 1990/91 and 1995/96. Taking into account process and distribution losses, it is estimated that 36 new nitrogen fertilizer complexes each based on a 1,350 tpd ammonia plant will be required to meet this need. The estimated total cost of these new plants and associated facilities is about US$6.7 billion. In addition, allowance must be made for replacing some of the old worn-out capacity. A detailed survey of existing ammonia capacity indicates a current closure rate of about 1.5 million tpy and this rate will most likely increase as the average age of existing capacity increases. Most existing ammonia capacity was built in the 1960s and 1970s, and in 1985, of the 81% of existing capacity that can be accounted for by age, about 6% was older than 20 years. By 1990 this will have grown to more than 20% and by 1995 to nearly 40%. Obviously a major reinvestment program to replace some of these older plants will be needed. However, even on the present replacement basis of 1.5 million tons per year, it is estimated that replacement capacity needs between 1990/91 and 1995/96 will amount to 20 new plants. In total therefore a minimum of 56 new plants with an estimated investment of more than $10.4 billion will be required between 1990/91 and 1995/96.

Nitrogen Fertilizer Prices

Both nitrogen fertilizer prices and ammonia prices have fluctuated widely over the years. The ratio of urea and ammonia prices has also fluctuated considerably indicating that to some extent the export markets for these materials are not always the same. In 1974 there was a major hike in fertilizer prices and urea and ammonia fob prices (1986 dollar basis) increased to US$530/ton and US$570/ton, respectively. Prices fell sharply in the next two years as new capacity came on-stream. There was a slight improvement in prices in 1981 but a further decline since then. Ammonia prices made a temporary recovery in 1984 probably due to increased demand in the USA in that year and after about 2 million tons of capacity in the USA had been permanently closed in the previous two years. Between 1984 and 1986, however, both ammonia and urea prices have declined steadily to reach very low levels.

Nitrogen fertilizer prices are expected to remain low during the next two years as a weak market in the USA and low agricultural prices depress demand. After 1988/89, however, as supply demand balances diminish, prices are

In the Centrally Planned Economies, Eastern Europe, mainly through the USSR and Romania, will have a very large nitrogen surplus on which the world will depend to maintain a balanced situation. China is expected to maintain its dependence on imports of about 2-2.5 million tons of N py.

Due to the relatively long lead time for nitrogen fertilizer projects to be implemented, it will be difficult to realize a significantly higher supply situation than that projected now for 1989/90. It seems highly likely that if total nitrogen demand continues to increase as projected, at about 2.4% py (a relatively pessimistic growth rate compared with previously projected rates), a tight situation will develop in two or three years' time. Unfortunately, very low prevailing prices, and worse prospects for the immediate future are inhibiting further investment in new nitrogen fertilizer plants both in Developed and Developing Market Economies. The USSR, because of its large surplus capacity, will continue to dominate the world export market for many years to come. If for one reason or another the USSR does not maintain this high level of exports, for example, by increasing its own consumption more rapidly, or by failing to bring all its plants to full production as forecast, the tight situation projected for 1989/90 will be further aggravated.

World Nitrogen Fertilizer Trade

Fertilizer trade in nitrogen products and ammonia has grown steadily in all regions in recent years. In the last two decades, trade in nitrogen fertilizers has grown from about 3 million tons to more than 14 million tons.

to have a major impact on production costs in this section of the nitrogen industry. If oil and gas prices remain low in the Developed Market Economies, there may be some new investment particularly on existing sites. If oil prices increase significantly it will accelerate the trend of new plants being erected where gas is very cheap. Crude oil prices would however need to rise to about \$30/BBL before other fuels such as fuel oil, naphtha and coal are likely to be competitive with natural gas-based plants.

World Fertilizer Nitrogen Consumption and Food Production

The projection of nitrogen fertilizer demand growth of 2.6% py through the next ten years is basically the result of aggregating country and regional forecasts of the World Bank/FAO/UNIDO/Industry Fertilizer Working Group. After three decades of rapid growth, nitrogen demand has fallen sharply but this has not as yet been reflected in any perceived fall-of in food production in the form of world grain production. Some exploratory statistical analysis has been made of grain and nitrogen fertilizer use time series to see how this compares with current nitrogen projections. Although the results are indicative only and far from conclusive the analysis shows that based on historic relationships between grain production, nitrogen fertilizer use and time, a growth rate of about 4.0% py in nitrogen fertilizer demand will be needed to continue the present trend in grain production which would improve the grain production per capita by 20% by the year 2000. The nitrogen growth rate to maintain food production per capita at the same rate in the year 2000 as it is in 1986 is estimated to be 2.7% py. It is recognized that there are many other factors improving the increase in grain production which cannot be fully accounted for, and it is believed that this particular issue requires further investigation.

Future Outlook for the World Nitrogen Industry

Some forecasts have been made of countries and regions where new nitrogen plants might be built to meet world nitrogen requirements in the period 1990-95. It is estimated that about 56 new large plants will be needed of which about 7 might go to the Developed Market Economies, 14 to the Centrally Planned Economies and the remaining 35 (about 60% of the total) to the Developing Market Economies. There will be opportunities for those countries with cheap natural gas, particularly in Latin America and the Far East, where there are opportunities for both domestic and export markets. There will also be opportunities for those countries in the Near East who have cheap gas and have already established an infrastructure for nitrogen exports. For some countries, however who may have cheap gas, the very large investment cost needed to establish a new fertilizer plant with the necessary gas supply and product distribution infrastructures, may be too high to ensure that such a plant will be capable of giving an adequate return on the investment.

I. INTRODUCTION

1.01 Nitrogen is an important element both in agriculture and the chemical industry. It is one of the three primary plant nutrients which comprise nitrogen, phosphorous and potassium. Nitrogen products are nearly all produced from ammonia which is derived from the catalytic fixation of atmospheric nitrogen. About 80-85% of all ammonia produced is used to manufacture nitrogenous fertilizers and the remainder is used in the chemical industry to produce a wide range of products such as textiles, explosives, animal feeds etc.

1.02 A significant part of new agricultural production in the future will come from increased nitrogen fertilizer use and so the availability and source of new ammonia production will continue to be vitally important to the world's agricultural industry. As the feedstock for ammonia is hydrocarbon in its various forms, although mainly as natural gas, both the availability and cost of energy are important in determining the economics of new ammonia production.

1.03 This World Nitrogen Survey is designed to give a comprehensive overview of the nitrogen industry, with particular emphasis on recent developments, present situation and projected future developments in demand up to 1996 and supply and balances up to 1991. Investment and production costs are also presented and some attempts have been made to establish relationships between fertilizer prices and available supply, and also, between food production and nitrogen fertilizer consumption.

1.04 The structure of the report is systematic and attempts to make each chapter self-contained. Extensive annex material on supply/demand and production and investment costs are given to support the global view of the report.

II. THE WORLD NITROGEN INDUSTRY

A. Historical Development

2.01 Nitrogen is the principal constituent of the earth's atmosphere, comprising 78% of its volume. It can be obtained from the air in two distinct forms, either in the fixed form with other elements, or in its unaltered elemental form. Nitrogen is used almost entirely in its fixed form and its most important role is as a primary crop nutrient. Fixed nitrogen is supplied in rainwater and atmospheric nitrogen is fixed by certain leguminous crops that are included in the crop rotational system. Until the beginning of this century, natural supplies of nitrogen from a system of crop rotation were deemed adequate to meet crop needs.

2.02 At the beginning of the century, a growing population placed increasing emphasis on the need to augment agricultural production by increasing nitrogen fertilizer application and considerable research was initiated to develop new processes for the fixation of atmospheric nitrogen into forms which could be used for the production of nitrogen fertilizers. As a result of this work, three new processes were developed to fix nitrogen.

2.03 In 1903, the "Arc" process was introduced commercially in Norway. Nitrogen and oxygen were combined to form nitric oxide at very high temperature and then further reacted to form nitrogen dioxide and nitric acid from which the fertilizer, calcium nitrate was produced. Another process was developed about the same time to produce calcium cyanamide. Calcium carbide produced by the reaction of lime with coke in an electric furnace was converted to calcium cyanamide by reaction with nitrogen extracted from the air. Both of these processes however were only economic where cheap electric power was available and because of this, were not widely adopted, although there are still a few plants operating today where cheap power is available. The most important development in nitrogen fixation was the Haber-Bosch process which in 1913 was introduced on a commercial scale in Germany. This process was based on the catalytic reaction between hydrogen and nitrogen at high temperature and pressure and present day ammonia plants still employ this basic process.

2.04 Up to about 1960, most of the world ammonia capacity was in the developed countries, such as the USA, West Europe and Japan. Plants were small and served local markets and a wide range of feedstocks was used. There were very few nitrogen plants in developing countries and only a small quantity of nitrogen fertilizers was imported by these countries. During the 1960s and 1970s, several major changes took place in the industry. In technology, the trend to large plants that could use centrifugal compressors had a major impact on reducing the cost of producing ammonia. Although the price of oil and naphtha rose steeply during the 1970s, generally, natural gas became more available and remained relatively cheap. The nitrogen industry expanded rapidly in the 1960s and 1970s as both fertilizer and industrial use of nitrogen increased. The structure of the industry also changed significantly as both production and consumption of nitrogen in the Centrally Planned Economies and Market Economies grew rapidly. In 1984/85 world nitrogen production and consumption for both fertilizer and industrial use were 86.6 and 80.8 [a/] million tons N respectively made up percentage-wise as follows.

	Developed Market Economies	Developing Market Economies	Centrally Planned Economies
Consumption (% of total)	38	20	42
Production (% of total)	41	17	42

a/ This production figure is based on downstream products. The corresponding ammonia production in 1984/85 is estimated at 87.6 million tons N.

The major producers were USSR, China, USA and India, and these countries were also the major consumers of nitrogen products. About 12% of all nitrogen produced went to industrial use, mainly in the Developed Market Economies. About 80% of ammonia plants were based on natural gas and more than 70% of the world's nitrogen industry was owned by the public sector.

B. The Fertilizer Nitrogen Industry

2.05 Information on nitrogen production in Figure 1 shows how the industry has grown since the beginning of the century and in particular during the last two decades. Up to about 1950, the nitrogen fertilizer industry developed relatively slowly and in fact lagged well behind the growth of the phosphate and potash fertilizer industries. The main nitrogen fertilizer for many years was ammonium sulphate, produced either from coke oven gas, by reacting ammonia with sulphuric acid, or from the reaction between gypsum, carbon dioxide and ammonia (Merseberg process).

2.06 The second most important nitrogen fertilizer was ammonium nitrate, particularly in Europe where it was the main component of nitrophosphate fertilizer, or as calcium ammonium nitrate. Other fertilizers included sodium nitrate, cyanamide and urea. In 1950, however, world urea production was less than 100,000 tpy and by 1960 was still only 550,000 tpy. In 1950 world production of nitrogen fertilizers was just over 4 million tons of N, of which 75-80% was produced in the Developed Market Economy Countries. Trade in nitrogen fertilizers was relatively small.

2.07 In the 1960s and 1970s there was a rapid increase in fertilizer use, particularly in developing countries. The "green revolution", with the increasing use of high yield varieties of cereals, accelerated the growth of nitrogen fertilizers, particularly urea. Growth rates for nitrogen fertilizer consumption world-wide between 1960 and 1980 averaged almost 10% per year. At the same time as nitrogen production and use were increasing rapidly, major changes in the structure of the industry were taking place. One of these was the increase in production facilities in developing countries for the manufacture of ammonia and urea. Many developing countries recognized the importance of building up a domestic fertilizer industry to support their agricultural programs. Urea manufacture, which can be conveniently integrated with ammonia production, represented one of the most economic methods of producing large quantities of nitrogen fertilizers, particularly for those countries with resources of natural gas.

2.08 As a result of these developments, the production of urea has grown rapidly in the last 25 years from virtually nothing, to make it one of the most important nitrogen fertilizers. In 1955 about 400,000 tons of N as Urea were used as a fertilizer representing about 5% of fertilizer use. In 1984/85, 37% of the 70 million tons of N used as fertilizer was as urea. The growth of urea production as a percentage of total nitrogen production is shown in Fig. 2. In addition to the major changes in developing countries, similar changes were taking place in the Centrally Planned Economies. In China, through the 1960s and 1970s, nitrogen fertilizer production and use increased at an incredible average rate of 30% per year. One other very important development has been the rapid increase in production facilities in the USSR during the last decade. About fifty new large-scale plants have been built, many with the help of western engineering companies which have allowed the USSR to become the largest producer and exporter of nitrogen products in the world. The growth of the world nitrogen fertilizer industry in the last 20 years is shown in Figures 3 and 4. Both production and consumption have grown at a higher rate in the Developing Market Economies and the Centrally Planned Economies than in the Developed Market Economies.

Figure 1:
WORLD NITROGEN PRODUCTION

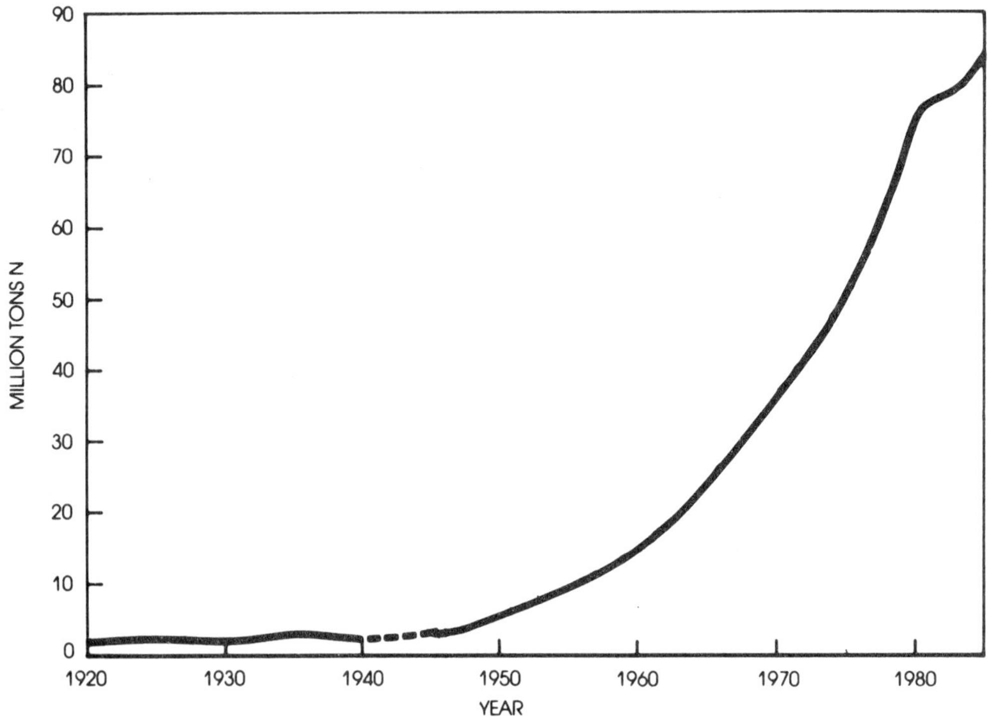

Figure 2:
UREA AS PERCENTAGE OF TOTAL NITROGEN PRODUCTION

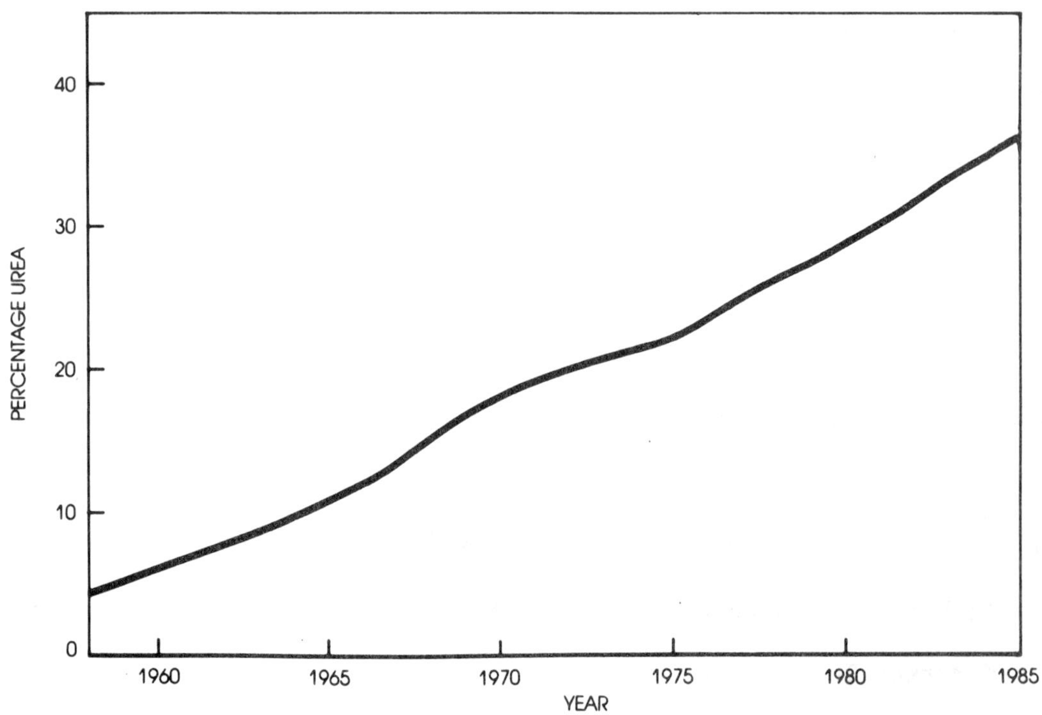

World Bank—30178:2

Figure 3:
NITROGEN FERTILIZER PRODUCTION

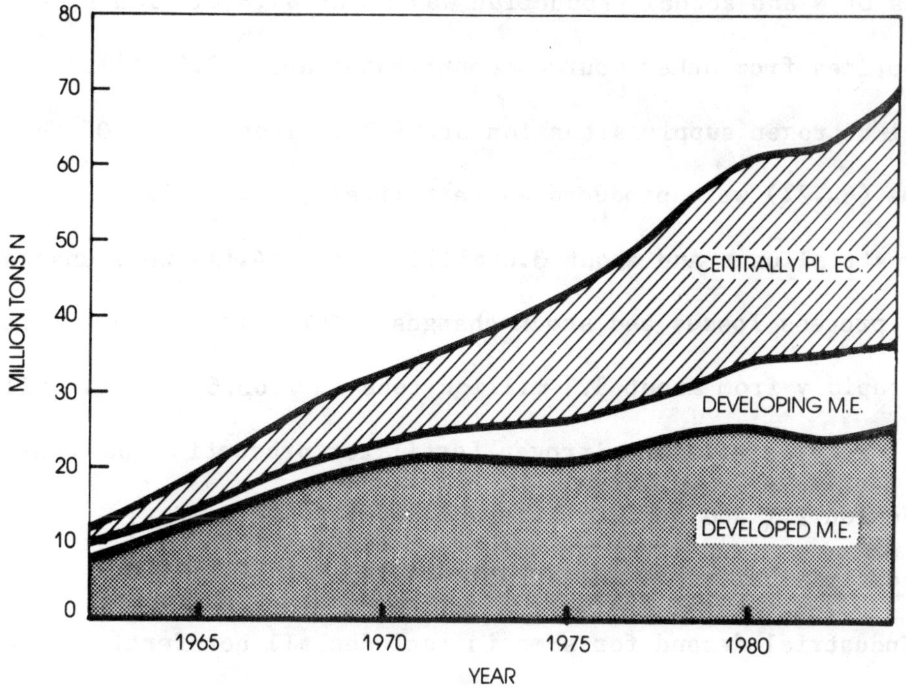

Figure 4:
NITROGEN FERTILIZER CONSUMPTION

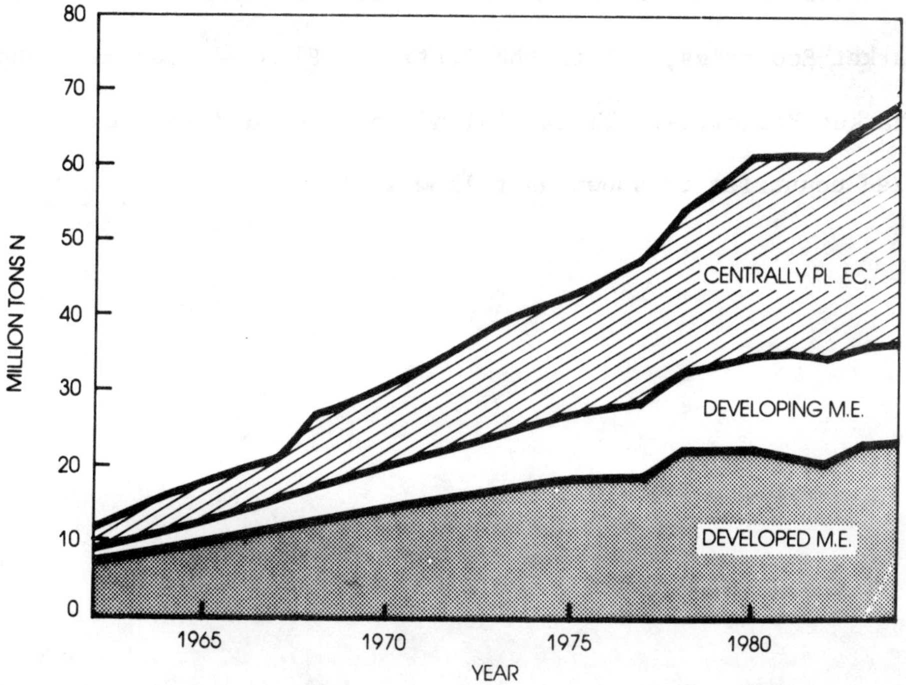

World Bank—30178:3

2.09 In 1984/85 world ammonia nameplate capacity was estimated to be 108 million tons of N and actual production was about 87.6 million tons of N. Nitrogen supplies from other sources contributed about 0.6 million tons of N to make a total nitrogen supply situation of 88.2 million tons. Of this, 74.3 million tons (84.2%) were produced as fertilizers, 10.3 million tons (11.7%) went to industrial uses and about 3.6 million tons (4.1%) were unaccounted for in terms of process losses and stock changes. Trade in nitrogen fertilizers has also grown rapidly from about 3.3 million tons in 1963/64 to 16.4 million tons in 1984/85. More details on nitrogen fertilizer production and consumption are given in Section III and on fertilizer trade in Section IV.

C. The Industrial Nitrogen Industry

2.10 Industrial demand for ammonia includes all non-fertilizer applications, such as fibers/resins/plastics/explosives, livestock feeds, and a variety of many other products. Country and regional industrial nitrogen demand is given in Annex 6. In total, in 1984/85, industrial nitrogen consumption was about 13% of total nitrogen use. About 68% of this demand occurred in the Developed Market Economies, 26% in the Centrally Planned Economies and 6% in the Developing Market Economies. Industrial nitrogen demand in the highly industrialized countries is shown as follows.

Table 1

Major Users of Industrial Nitrogen 1984/85

Country	Industrial N Use ('000 tons N)	Industrial N as Percentage of Total N Use
Japan	1,190	63
Germany (FR)	653	31
USA	2,700	21
UK	370	19
France	433	16
USSR	1,390	12

By contrast, industrial nitrogen consumption in major developing countries such as China and India is believed to be less than 5% of total nitrogen use.

2.13 Apart from the USA, little information is available on the breakdown of industrial nitrogen use, but based on information from USA statistics, it is estimated that about 50% of industrial demand goes into fibers, plastics, etc., about 15% into explosives, 10% into livestock feed and about 25% into other uses. One of the main nitrogen intermediates in industry is nitric acid, which is used to manufacture ammonium nitrate (an explosive), adipic acid (used in the manufacture of nylon) and for many other important chemicals, such as isocyanates and nitrobenzene. Apart from its fertilizer use, urea is also an important chemical ingredient for the manufacture of urea formaldehyde resins and melamine, as well as an animal feed and about 10% of urea production in 1984/85 went to non-fertilizer uses. Ammonia is also used directly for the production of some industrial chemicals, such as acrylonitrile, hexamethylenediamine and also in the production of ammonium phosphates used as an animal feed and in fire control products.

2.14 It is possible that a certain amount of ammonia currently unaccounted for as losses, stock changes, etc., goes unrecorded into a variety of small chemical products.

III. REGIONAL AND WORLD NITROGEN SUPPLY-DEMAND BALANCES

A. General

3.01 Regional and world supply demand balances are based on those prepared by the World Bank/FAO/UNIDO/Industry Fertilizer Working Group June 1986. This Group maintains a comprehensive data base on fertilizer plant capacity and fertilizer consumption and meets twice a year to update this information. Industry members of the Group represent the International Fertilizer Association and several major trade associations from the nitrogen, phosphate, potash and sulphur industries in both developed and developing countries.

3.02 As almost all nitrogen is initially produced as ammonia the capacity for nitrogen is based on the world capacity of ammonia plants to which is added the small amount of nitrogen derived from non-ammonia sources. This includes nitrogen recovered from coke ovens, synthetic textile production, cyanamide and sodium nitrate. Country ammonia capacities have been derived from the Fertilizer Working Group's plant lists and the supply capability is calculated from the capacities using special phasing in and out factors for changes in capacity. An average plant utilization rate is assumed year by year for each country which takes into account historical and anticipated future performance. Approximately 85-90% of all ammonia produced goes to fertilizer use and the remainder to a variety of industrial uses. Fertilizer ammonia is estimated by deducting industrial use from the total supply. Further allowances, to take into account process and distribution losses, are then made on a regional basis. The losses are based on historical information and are reviewed periodically. Finally, the nitrogen balances are estimated by subtracting the regional fertilizer demand from the fertilizer nitrogen supply availability.

Regional demand is assessed by the Fertilizer Working Group on a "Delphi" principle but individual contributions are based on a range of methodology including trend projections, econometric modelling, agricultural programs and sales forecasts. The regional and world supply demand balances are given in Annex 7.

B. Developed Market Economies

3.03 Overall, little change is expected in the supply/demand balances in the Developed Market Economies between 1985/86 and 1990/91. Over this period, ammonia capacity will increase slightly by about 4% and the overall growth in nitrogen fertilizer demand will follow a similar trend. Due to a depressed agricultural situation in the next two or three years, overall demand is expected to decline before starting a slow recovery again in 1987/88.

1. North America

3.04 The North American situation is dominated by the US situation which is covered in more detail in the Country Profile section in Annex 20. It is estimated that in 1984/85 the overall supply/demand situation in North America was finally balanced. Most of the US idle capacity which had not been permanently closed in the previous two years was back on stream and operating near 90%, although about three million tons of the 5 million tons of capacity which was idled between 1980 and 1985 has closed permanently. There is no known planned capacity increase in the USA before 1990 but several new plants now under construction or planned in Canada will increase capacity there by about one million tons. The US Food Security Act of 1985, also generally known as the "New Farm Bill" is certain to have a major impact on nitrogen fertilizer consumption in the USA with reductions of 7% or more in 1986. A further fall in consumption is also likely in 1987 but hopefully thereafter consumption will start to increase again and by 1990/91 reach the level achieved in 1984/85.

With declining demand and new capacity coming on stream in Canada, there will be a net surplus of nitrogen in the region through the remainder of the decade.

2. Western Europe

3.05 Although the ammonia supply potential in the region is almost sufficient to meet its requirements, there is a net import of ammonia of more than 2 million tons to meet nitrogen fertilizer export needs. There is considerable interregional trade as well as regional trade in both ammonia and nitrogen products, as many countries in Western Europe at the present time, with low export prices, find it cheaper to import rather than produce ammonia domestically. A year or two ago it looked as if there would be little new capacity in Western Europe but the more favorable outlook for world energy prices may modify this situation and several new plants planned to come on stream at the end of the decade have recently been announced. With a depressed situation in the EEC, little growth is expected in fertilizer consumption. Through 1990/91 the growth rate is projected at about 1.2% py and through 1995/96 about 1% py. Most of this will occur in the Mediterranean countries where the application rates are still relatively small. Little change is expected in the regional supply/demand balances through the remainder of the decade.

3. Oceania

3.06 This region comprising mainly Australia and New Zealand is not a major producer or consumer of nitrogen fertilizer. A new nitrogen project is being considered in Western Australia but seems unlikely to come on stream before the end of the decade. With a change to more arable farming in Australia, nitrogen demand is expected to increase at about 2.2% py and the region, already a mall net importer of N will have to increase its imports by a further 50,000 tons of N before new domestic capacity is available.

4. Other Developed Market Economies

3.07 This economic classification has no geographical significance and includes Israel, Japan and South Africa. The only significant change will take place in Japan where excess capacity will close under a Government rationalization program.

C. Developing Market Economies

3.08 A significant improvement in the supply situation of Developing Market Economies will take place in the next five years. Installed capacity will increase by about 50% including increases in all regions but particularly in the Far East. Even so, with an increasing demand of nearly 6% py the region will still remain a significant importer of nitrogen fertilizers at the end of the decade.

1. Africa

3.09 The relatively small existing ammonia capacity in Africa will increase from about 0.4 million tons to 1.3 million tons by 1990/91 due mainly to increases in Algeria and Nigeria. The current nitrogen fertilizer use in Africa is very low and although it is forecast to increase by about 5.9% py, it will still be well below the needs in the foreseeable future. The region will remain an importer of nitrogen fertilizer through 1990/91.

2. Latin America

3.10 Latin America has a large potential to increase its capacity as well as demand and considerably growth is expected in the region during the next five years. Mexico and Trinidad are the main producers based on large gas reserves and a considerable expansion of ammonia capacity in these countries is planned during the next five years. Capacity in the region will increase by about 25% from 5.8 to 7.2 million tons. Growth in nitrogen fertilizer demand in the region will take place mainly in Mexico and Brazil and the overall yearly growth rate in demand will average 6.1% over the next five years during which time some of the consumption which was lost in the early 1980s will be recovered. The current surplus in the area of about 0.8 million tons N is expected to decline sharply until new capacity comes on stream at the end of the decade.

3. Near East

3.11 With large resources of natural gas, this region will also increase its capacity by about 50% during the next five years but the major benefits of this new capacity will not be fully realized until the early 1990s when the capacity is completely phased-in. Demand is forecast to grow at about 3.9% py and the region will become an increasing exporter of nitrogen with a surplus of about 1.0 million tpy by 1990/91.

4. Far East

3.12 The Far East has been one of the major importers of nitrogen fertilizers for many years due mainly to the increasing needs of India, Indonesia, Pakistan, Bangladesh, etc. Within the last few years however most of these countries have increased their capacity significantly and this expansion will continue throughout the 1980s. Consequently, the region will increase capacity by about 5.1 million tons of which about 3.2 million tons is planned to come on-stream in India. Demand, however, is expected to increase at about 6.1% on average per year and this will more than offset the increasing supply capability and so the deficit of more than two million tons will continue through the end of the decade.

D. Centrally Planned Economies

3.13 The two main regions within the Centrally Planned Economies are Eastern Europe and Socialist Asia.

1. Eastern Europe

3.14 The centrally planned region of Eastern Europe also includes the USSR which is by far the largest producer. After a sharp rise in new capacity in the early 1980s only about 0.5 million tons of additional new plant will be commissioned through 1990/91. Supply capability will however rise by about 1.5 million tons as much of the capacity commissioned at the beginning of the 1980s is phased in completely. Demand in the region is expected to increase at about 2.6% py on average throughout the remainder of the 1980s. Eastern Europe and particularly the USSR have a large surplus of nitrogen which is currently exported. Russia however is also a large importer of food and if there is a significant change in government policy regarding increased domestic production of food using nitrogen fertilizer, this could reduce the surplus in the USSR and hence the amount available for export. The surplus will also depend to a large extent on the ability of the USSR to commission fully and improve the capacity of existing plants. Although not specifically included in the forecasts, it is reported from Russian sources that about 1 million tons of ammonia capacity will be closed by the end of the decade. Whether or not this will be replaced by new plant is not known at this stage.

2. Socialist Asia

3.15 In this region, fertilizer statistics are dominated by those from the People's Republic of China both with regard to production and demand. Mainland China accounts for about 93% of the total regional nitrogen fertilizer demand.

Capacity in Socialist Asia will increase by about 1.5 million tons over the next five years as a large investment program in China is completed. Demand for nitrogen fertilizers in Socialist Asia is forecast to increase at about 2.9% py in the next 5 years leaving a yearly deficit on average of about 2-3 million tons of nitrogen which will need to be filled by imports. These forecasts are based on the assumption that China will concentrate on increasing use of other nutrients to correct its present nutrient imbalance.

E. World Ammonia Capacity and Supply Capability

3.16 World ammonia capacities and utilizations since 1970 with forecast capacities and potential utilizations for the next five years are given below. As little informations is available on the phasing-in performance of older plants, nominal or name-plate capacity has been used throughout which gives a slightly lower figure than for effective capacity.

Table 2

World Ammonia Capacity and Supply Capability 1970-1990

Year	Nominal Capacity (Million tons N)	Ammonia Production a/ (Million tons N)	Utilization %
1970/71	56.7	43.2	77.6
1973/74	64.6	53.2	82.3
1977/78	84.8	62.9	74.2
1980/81	95.7	76.5	79.9
1983/84	107.4	84.1	78.3
1984/85	108.0	87.6	81.1
1985/86	111.7	91.0	81.5
1986/87	113.5	93.6	82.5
1987/88	115.9	96.5	83.3
1988/89	118.0	98.4	83.4
1989/90	120.4	100.2	83.2
1990/91	122.5	101.9	83.2

a/ After 1984/85 figures refer to potential supply.

Potential supply capability takes into account the technical constraints to production, based on historical performance, location of plant and other factors. It allows for the time to phase in new capacity and also for idle plant. It assumes there are no market constraints to production. The figures in Table 2 contain some allowance for idle plants in 1983-1985 but assume that all idle plants will be available in 1986/87 and thereafter. Potential utilization will increase very slightly over the next few years as plants now being commissioned reach full production. The ammonia capacities in Table 3 show how the structure of the industry has changed over the last two decades as countries in both Developing Market Economies and Centrally Planned Economies have increased capacity.

Table 3

Major Countries Ammonia Capacities (Million tons N per year)

Country	1975/76	1980/81	1985/86 (Estimated)	1990/91 (Projected)
USSR	11.6	18.2	24.5	24.9
USA	14.2	15.6	13.6	13.7
China	6.0 a/	16.0	16.9	18.3
India	2.6	4.8	7.0	9.8
Romania	2.1	3.1	3.5	3.5
Germany, F.R.	2.4	2.4	2.0	2.0
France	2.5	2.8	2.5	2.8
Netherlands	2.3	2.2	2.7	2.7
UK	1.4	1.7	1.8	1.9

a/ Estimated.

In the last decade the USSR has overtaken the USA in ammonia capacity and is now the world's largest producer. China is the world's second largest producer of ammonia and nitrogen fertilizers. On the other hand, ammonia capacity in most European countriea has remained relatively stable in the last few years and seems likely to continue so through 1990.

3.17 Although considerable new capacity will continue to come on-stream world-wide during the five year period 1985-1989, as can be seen in Annex 9 and

Table 4, commissioning and construction of new ammonia capacity will fall off considerably during this period compared with previous 5-year periods since 1965. Taking into account the lead time to plan, finance and build new large ammonia plants, particularly in developing countries it is unlikely that there will be much change in this situation and any new facilities that can be realized will probably be offset by delays in other plants already planned.

Table 4

Ammonia Plants Commissioned within 5-Year Time Periods 1960-1990

5-Year Span	1960-64	1965-69	1970-74	1975-79	1980-84	1985-89 b/
Developed Market Ec.						
No. New Plants	19	63	41	41	28	11
Million tons Capacity	2.14	10.98	9.03	8.71	3.86	2.66
Developing Market Ec.						
No. New Plants	9	20	25	34	29	32
Million tons Capacity	0.66	2.14	5.13	5.78	6.76	7.90
Centrally Planned Ec. a/						
No. New Plants	21	32	40	46	30	12
Million tons Capacity	2.19	5.09	7.06	13.09	7.76	3.68
World Total						
No. New Plants	49	115	106	121	87	55
Million tons Capacity	4.99	18.21	21.22	27.58	18.38	14.24
Average Plant Size (000 tpy)	102	158	200	228	211	259

a/ Excludes small Chinese plants
b/ Estimated future capacity

F. World Nitrogen Demand

3.18 Total world nitrogen consumption is forecast to increase from 80.8 million tons of N in 1984/85 to 94.4 million tons in 1990/91, an average yearly increase of 2.6%. Between 1990/91 and 1995/96 when total demand is projected to be 104.6 million tons, the average increase is 2.2% py

1. World Fertilizer Nitrogen Demand

3.19 Table 5 indicates how fertilizer nitrogen demand has increased over the last thirty years. Regional and major country consumptions over the last decade are given in Annex 4. World fertilizer nitrogen consumption has grown steadily over the last three decades, increasing more than 10 fold. There have been two major recessions in demand, the first took place in 1975 due to a major price hike during the so-called "fertilizer crisis". The second occurred in early 1980s as a result of the global recession and the PIK program in the USA. Average yearly increases in fertilizer nitrogen demand over successive five-year periods and forecasts for the next two five-year periods since 1984/85 are shown below.

Table 5

World Nitrogen Fertilizer Consumption 1955-1995 (Million Tons N)

Year	54/55	59/60	64/65	69/70	74/75	79/80	84/85	89/90 a/	94/95 a/
Consumption	6.2	9.2	15.3	28.7	38.6	57.3	70.5	80.4	90.7
5 year increase		3.0	6.1	13.4	9.9	18.7	13.2	9.9	10.9
% yearly increase over 5 year period		8.2	10.7	13.4	6.1	8.2	4.1	2.7	2.4

a/ Forecast demand.

Between 1984/85 and 1995/96 when fertilizer demand is expected to reach 93.3 million tons, demand is expected to increase on average at 2.6% py.

2. World Industrial Nitrogen Demand

3.20 Industrial use of nitrogen up to about 1970, was growing more quickly than fertilizer use but afterwards fertilizer demand grew more quickly. World industrial nitrogen consumption is forecast to increase at about 2.0% py from 1984/85 through 1990/91. In the Developed Economies, it will be 1.0% py, in the Developing Economies 3.8% py and in the Centrally Planned Economies 2.1% py. Detailed projections are given in Annex 6.

G. World Nitrogen Supply-Demand Balances

3.21 The detailed balances are given in Annex 7 and are summarized in Table 6.

Table 6

World Nitrogen Balances
(Million metric tons of N)

	1985/86	1986/87	1987/88	1988/89	1989/90	1990/91
Developed Market Econ.	-0.20	0.52	0.81	0.60	-0.26	-0.40
North America	0.33	1.13	1.66	1.63	0.75	0.60
Western Europe	-0.36	-0.19	-0.37	-0.50	-0.42	-0.36
Oceania	-0.03	-0.04	-0.05	-0.06	-0.08	-0.09
Other Developed M.E.	-0.14	-0.39	-0.42	-0.47	-0.50	-0.55
Developing Market Econ.	-1.44	-1.41	-1.38	-1.57	-1.53	-1.48
Africa	-0.60	-0.64	-0.49	-0.43	-0.38	-0.40
Latin America	0.97	0.70	0.42	0.24	0.20	0.25
Near East	0.40	0.73	0.79	0.91	0.92	1.01
Far East	-2.22	-2.20	-2.11	-2.29	-2.28	-2.35
Centrally Planned Econ.	4.97	4.74	4.38	3.71	2.58	1.37
Socialist Asia	-2.20	-2.31	-2.27	-2.33	-2.75	-3.22
Eastern Europe	7.18	7.04	6.65	6.05	5.33	4.60
World	3.33	3.84	3.81	2.74	0.79	-0.51

3.22 The world fertilizer industry is well known for its cyclical price behavior and supply demand balances have fluctuated considerably through the years. In the late 1960s and early 1970s there was a period of surplus but this turned to a short supply situation in 1974. This situation was exacerbated by the increased buying which took place in face of anticipated fertilizer shortages. The rapid increase in prices of nitrogen fertilizer slowed down nitrogen demand but at the same time stimulated an increase in new ammonia capacity. The result was that there was a further oversupply situation in 1977/78. A surge in nitrogen demand in 1979 however once more resulted in a more balanced situation, with an improvement in nitrogen fertilizer prices. In the following years, due to the global recession and other factors, both fertilizer and industrial nitrogen demand stagnated. At the same time considerable new nitrogen supply became available, as many new ammonia plants with export potential came on-stream particularly in Eastern Europe. In a situation of surplus supply and low prices several million tons of capacity, mainly in the USA, became idled or closed permanently. In 1983/84 fertilizer demand improved significantly and nitrogen use increased by about 10% worldwide and much of the idle capacity came back on-stream. In 1984/85 nitrogen fertilizer demand increased by 5.0% and although this was offset by an increase in supply capability, a more balanced situation was achieved and nitrogen fertilizer prices improved. In 1985/86, however, with a fall-off in world demand due mainly to the depressed agriculture situation in the USA and Western Europe, the surplus has increased and prices have fallen.

3.23 The World Bank/FAO/UNIDO regional nitrogen fertilizer demand projections through 1995/96 are shown in Annex 5. Between 1984/85 and 1990/91 world fertilizer nitrogen consumption is forecast to increase on average at 2.7% per year and over the five following years at an average yearly rate of 2.4%.

3.24 The balances in Table 6 indicate no major changes in the structure of the nitrogen fertilizer market. The Developed Market Economies will more or less be able to meet their own demand but will continue to take advantage of imports of cheap material. The Developing Market Economies will remain major importers to meet their needs well into the early 1990's. The Centrally Planned Economies will maintain a large surplus situation. Overall, the world supply surplus will continue to increase through 1987/88. Thereafter, the balances will decline and in 1990 a balanced situation will result. Unless additional plants come on-stream or nitrogen demand falls below its projected average increase of 2.7%, a tight market situation may develop. The extent to which this happens will depend on several factors which are discussed below.

3.25 In a situation of excess supply and depressed agricultural conditions fertilizer nitrogen prices in real terms have slumped to their lowest levels ever. In order to justify new investment, producers will need to anticipate much higher prices. Few are prepared to do this at the present and this has resulted in a slow-down in new investments particularly in the Developed Market Economies.

3.26 The US nitrogen industry suffered a major shock in the early 1980s when in the face of very cheap ammonia imports, many domestic plants could not compete and were forced to close down. The uncertainty of natural gas prices, particularly with the recent de-regulation, has also made producers very cautious about further investment and some are looking at overseas sources, where gas is cheaper, to supplement their needs. As a result, there is currently no major new capacity planned in the USA during the next five years. With a deteriorating domestic market in the next two years, there will be even less inclination to invest, even though the more recent prospects for gas prices look more promising.

3.27 Although there is considerable interest in investment in developing countries and major growth is expected in this region, particularly where gas is cheap, development is lagging behind expectations. One reason is that whereas the large growth in new ammonia capacity during the 1970s was for import substitution, during the 1980s, many new plants in developing countries are looking at export markets. Such plants are more difficult to justify in view of current low export prices and many countries already with major debts are also finding it difficult to finance these projects. In some Near East countries the reduction in associated gas availability, because of reduced oil consumption, is constraining new investment. The recent fall in oil prices (and hence related gas prices) has also reduced the advantages of some export projects in developing countries.

3.28 As the USSR now dominates world nitrogen production and exports, the situation in that country over the next few years will have a major impact elsewhere. Unfortunately, information in the USSR nitrogen industry is very limited so one can only speculate about its future. However, it should be appreciated that if the USSR for one reason or another does not maintain a high level of exports, for example, by increasing its ownconsumption more rapidly or by failing to bring all its plants to full production as forecast, the tight situation on world supply/demand balances forecast in 1990/91 could be further aggravated.

3.29 Considering the investments that were made in the late 1970s and current prices, Eastern European countries will carefully re-examine investment in this sector and in particular the opportunities of selling gas directly to other European Countries rather than a feedstock for fertilizers.

3.30 To summarize the overall situation on world supply/demand balances, it would seem that a tight supply situation could develop at the end of the 1980s. A short-term excess supply situation in the next two years with low prices, however, will hide the changing situation. New investment in ammonia capacity has fallen off recently and may continue to be constrained in the next few years because of the uncertainty of energy prices in the Developed Market Economies and because of the financial problems in many developing countries.

IV. WORLD NITROGEN FERTILIZER TRADE

A. General

4.01 The last twenty-five years have seen a major increase in both production and consumption of nitrogen fertilizers in almost all regions of the world. The fact that these have not always increased at the same rate has resulted in a steady increase in international trade in nitrogen fertilizers and intermediates. In 1959/60 exports of nitrogen fertilizers totalled about 2.56 million tons and by 1984/85, exports had increased to 16.4 million tons (about 22% of production). In 1959/60 about 70% of nitrogen fertilizer trade was from Western Europe. Elsewhere in the world at that time the principal exporters were Japan, Chile, USA and Canada.

4.02 The regional percentage share of nitrogen fertilizer trade in the last two years is shown in Table 7, which indicates its wide distribution. Many large countries or regions are both large exporters and importers of fertilizers, particularly Western Europe and North America. In the last two decades exports of nitrogen fertilizers outside the Developed Market Economies have increased from virtually zero to about 40% of all export trade in 1984/85. This is due mainly to new export based nitrogen plants in the Near East and Eastern Europe including the USSR. The major countries involved in fertilizer nitrogen trade are given in Table 8.

Table 7

Share Distribution of Nitrogen Fertilizer Trade

	Imports % Share			Exports % Share		
	1982/83	1983/84	1984/85	1982/83	1983/84	1985/86
Developed Market Ec.	49.6	50.4	44.1	59.5	56.4	58.2
N. America	20.1	24.9	21.5	23.5	22.6	27.0
W. Europe	26.9	23.9	21.2	32.9	31.5	28.5
Oceania	0.8	0.8	0.9	-	0.2	0.2
Other Developed Ec.	1.8	0.8	0.6	3.1	2.1	2.5
Developing Market Ec.	30.6	28.0	27.2	34.9	14.6	16.0
Africa	4.1	3.3	3.3	0.8	0.8	0.6
Latin America	8.3	6.2	6.3	1.8	2.2	2.0
Near East	7.0	7.2	6.4	6.2	7.6	10.0
Far East	11.2	10.5	18.9	2.8	4.0	3.4
Centrally Planned Ec.	19.8	22.4	21.0	28.9	29.0	25.8
Socialist Asia	15.7	18.3	18.3	0.3	0.3	0.2
E. Europe & USSR	4.1	4.1	2.7	28.6	28.7	25.6
Total	100%	100%	100%	100%	100%	100%
Trade million tons N	12.871	15.105	16.297	12.174	13.718	16.459

Source: FAO Statistics 1985.

Table 8

Country Trade in Nitrogen Fertilizers ('000 tons N)

Countries	Imports		Countries	Exports	
	1983/84	1984/85		1983/84	1984/85
USA	3,574	3,312	USA	1,886	2,932
China[a]/	2,327	2,900	USSR	1,676	1,878
Germany, F.R.	684	673	Netherlands	1,329	1,493
India	656	2,008	Romania	1,300	1,303
Iran	572	471	Canada	1,217	1,511
Turkey	257	294	Belgium	735	705
Vietnam	338	293	Germany, F.R.	508	521
Thailand	236	243	Norway	380	400
Philippines	190	178	Italy	334	459
Cuba	191	195	Qatar	330	332
Malaysia	190	194	Germany, D.R.	326	219

a/ Figures are for Mainland China only.

Source: FAO statistics.

B. Nitrogen Fertilizer Products Traded

4.03 The types of nitrogen fertilizer products traded have changed a great deal in the last two decades as indicated in the next table. The most important aspect of this change has been the increase in urea to become the most important fertilizer traded. Urea is a very convenient material to produce, as it can be economically integrated with ammonia production. It is a concentrated material and is relatively easy to transport with none of the hazards associated with ammonium nitrate. Producers looking at the export market and with cheap natural gas have tended to invest in urea production facilities. At the same time, a rapid rise in fertilizer consumption in developing countries and particularly in the Far East, where urea is an excellent fertilizer for wet paddy, has resulted in a major increase in urea trade. Although many developing countries, such as India, China, etc., have greatly increased their own domestic urea capacity, they are still very dependent on imports.

Table 9

Exports of Nitrogen Fertilizers by Product ('000 tons N)

	1962/63		1983/84	
Ammonium Sulphate	1,101	(35%)	1,001	(7%)
Ammonium Nitrate[a/]	692	(22%)	808	(6%)
Calcium Ammonium Nitrate and Calcium Nitrate	252	(8%)	1,392	(10%)
Urea	566	(18%)	6,603	(48%)
Others	534	(17%)	3,914	(29%)
Total	3,145	(100%)	13,718	(100%)

a/ Includes ammonium sulphate nitrate.

4.04 Although ammonium sulphate, produced from sulphuric acid and ammonia, was once a very popular fertilizer, it's use has now declined due to high cost of production and low analysis. It is now mainly recovered as a by-product from the textile industry. Its major source is therefore in the developed countries of Western Europe and Japan. Although its major markets are in the developed countries where it is produced, for traditional market reasons and also because on certain soils and crops ammonium sulphate claims a premium price, there are still a few markets for it in the Far East and Latin America.

4.05 Ammonium nitrate is produced mainly in Western Europe where it is the principal source of nitrogen fertilizers, usually as a component of NPK mixtures. Most trade in ammonium nitrate takes place as interregional trade in Western European countries but some is also exported to other regions. Ammonium nitrate is also used as an explosive and some trade statistics may include material for this purpose also.

4.06 A good part of the production of calcium ammonium nitrate is concomitant with nitrophosphate manufacture, mainly in Western Europe. Because of the relatively low concentration of this material (21-27% N), it does not form a major part of long distance trade.

4.07 As mentioned earlier, because of its favorable economics of production and high concentration, urea has become the most important nitrogen fertilizer export material. With a number of new export-based plants under construction in the Middle East and also where the quantity of urea exports and its percentage of the export market are likely to increase even further. The major urea exporters are the USSR, Romania, The Netherlands, and Bulgaria. The USA is both a large importer and exporter of urea but a major part of its urea exports is to the Far East from Alaska. The major importers of urea are China, India, USA and Iran.

4.08 A significant part of nitrogen shipped as compound fertilizers is as nitrogen in diammonium phosphate (DAP). Total diammonium phosphate shipped in 1984 exceeded ten million tons product. This is equivalent to about 1.8 million tons of nitrogen and of this more than 75% came from the USA. Much of the ammonia required for the production of DAP in the USA was imported.

C. World Ammonia Trade

4.09 Before 1960 there was virtually no trade in ammonia but the changing structure of the industry, which included the deep sea shipment of concentrated intermediates from locations where the raw materials were cheap, stimulated world ammonia trade. In the late 1960s there was a rapid increase in trade mainly to serve increasing needs in Europe. Transatlantic trade to bring the cheap ammonia from the new large plants in the USA was accompanied by considerable interregional trade in Europe. By 1980 about one million tons per year of transatlantic trade in ammonia had developed. During the early and mid-1970s trade developed more slowly, but after 1978, following the commissioning of much new capacity in USSR and the Caribbean area, there was a further large increase in ammonia trade. Although the early 1980s showed a slight fall due to depressed economic conditions, there was a sharp rise again in the next few years with ammonia trade reaching about 7.3 million tons (N) or nearly 10% of total world production by 1984. Details of world ammonia trade for 1984 are given in Annex 10 and summarized in Table 10. The USA and Western Europe are the two main markets accounting for nearly 80% of total world trade. Most of the remainder goes to India, South Korea, Tunisia, Jordan, etc., where it helps to support domestic fertilizer industries. The USSR has developed as the major ammonia exporter with 31% of all ammonia exports in 1984. The Caribbean urea, through Trinidad, Mexico and Venezuela, accounted for about 21% of exports.

Table 10

World Ammonia Trade 1984 ('000 Tons N)

Major Exporters 1984		Major Importers 1984	
Economy	Trade	Economy	Trade
USSR	2,351	USA	2,366
Trinidad	966	Spain	523
Canada	684	Turkey	471
Netherlands	539	Belgium	432
USA	397	Denmark	350
Mexico	406	Korea, Rep. of	331
Germany, F.R.	292	India	328
France	205	Finland	297
UK	196	UK	272
Venezuela	166	Germany, F.R.	232
Qatar	165	Sweden	190
Indonesia	145	Italy	181
Libya	134	Greece	167
Austria	87	France	150
Eire	71	Tunisia	133
Italy	67	Yugoslavia	99
Abu Dhabi	65	Taiwan	81
Norway	65		

Source: British Sulphur Corporation.

D. World Urea Trade

4.10 As discussed in section II-B, the production of urea has grown rapidly in the past two decades and its importance as a traded fertilizer has also grown accordingly. In 1973 about 3.5 million tons of N as urea were traded and in 1984 this figure had more than doubled to 7.6 million tons. Throughout this period China and India have been the main importers accounting for about 40% of imports. Urea trade has followed the fluctuations of the industry generally, falling in 1976 and again in 1981 and 1982 with the world agricultural recession. Over many years, China and India have retained their relative position as importers and the USA has increased its share of imports whereas Latin America's share has declined.

4.11 On the supply side there have been some major changes in market structure. Japan, which in 1973 had more than 30% of exports, in 1984 had about 1% due to the fact that with expensive feedstocks it can no longer compete with urea based on cheap gas. For similar reasons West Europe's share of the market has halved. As a result of large investments in nitrogenous fertilizer plants in the late 1970s, more than 40% of exports now come from East Europe - from Romania, Bulgaria and USSR.

Table 11

Urea Exporters' Market Share (%)

	1973	1984
West Europe	26.8	16.6
East Europe, excluding USSR	18.6	20.4
USSR	4.1	19.7
North America	7.0	13.1
Japan	31.9	1.5
Asia (excluding Japan)	10.1	19.5
Others	1.5	9.2
Total	100.0	100.0

World urea trade figures are given in Annex 11 and summarized in Table 12.

Table 12

World Urea Trade ('000 tons product)

Major Exporters			Major Importers		
Country	Trade		Country	Trade	
	1983	1984		1983	1984
USSR	2,847	3,286	China	4,263	4,389
Romania	2,008	1,573	USA	1,639	1,917
USA	997	1,151	India	1,055	3,379
Netherlands	988	1,118	Iran	839	639
Qatar	716	714	Vietnam	579	541
Canada	665	1,024	Hungary	397	397
Germany, D.R.	641	493	Philippines	335	319
Kuwait	523	588	France	281	310
Bulgaria	492	533	Turkey	218	206
Saudi Arabia	362	709	Cuba	214	272
Hungary	336	355	Malaysia	200	268
Libya	303	673	Sudan	182	84
Indonesia	301	199	Colombia	178	280

Source; Fertecon.

E. World Calcium Ammonium Nitrate Trade

4.12 Calcium ammonium nitrate (CAN) can be produced as a by-product from nitrophosphate processes. It is also produced by adding powdered limestone to ammonium nitrate as a means of reducing the hazard of handling and storing ammonium nitrate. World calcium ammonium nitrate trade figures are given in Annex 12 and summarized in Table 13. It will be noted that almost all CAN trade is interregional trade in Western Europe.

Table 13

World Calcium Ammonium Nitrate Trade ('000 tons product)

Major Exporters			Major Importers		
Country	Trade		Country	Trade	
	1983	1984		1983	1984
Netherlands	1,106	1,265	Germany, F.R.	2210	2,093
Belgium	747	742	Turkey	687	451
Romania	701	398	Netherlands	526	509
Germany, F.R.	509	594	Belgium	307	407
Austria	346	339	France	268	244
France	278	267	UK	196	293
Sweden	163	128	Austria	138	63

Source: Fertecon.

F. World Ammonium Sulphate Trade

4.13 Because its major source today is as a by-product, ammonium sulphate supply depends mainly on the economic situation in the textile industry rather than the fertilizer industry. At one time, ammonium sulphate was the major form of nitrogen fertilizer but due to the high cost of sulphuric acid and processing costs, it is not now normally synthesized directly. Ammonium sulphate exports have fallen from about 35% of total world trade to 6% between 1963 and 1984. Trade statistics are given in Annex 13 and summarized in Table 14.

Table 14

World Ammonium Sulphate Trade 1983 ('000 tons product)

Major Exporters			Major Importers		
Country	Trade		Country	Trade	
	1983	1984		1983	1984
USSR	1,035	965	Thailand	459	466
USA	660	611	Brazil	442	576
Belgium	636	696	Cuba	407	390
Italy	508	321	Turkey	288	410
Netherlands	443	415	France	261	305
Germany F.R.	307	303	USA	249	279

Source: Fertecon.

G. The Outlook for Future Nitrogen Fertilizer Trade

4.14 It is expected that the increase in fertilizer trade will be mainly as ammonia and urea, and so the discussion is limited to these two materials. With the exception of diammonium phosphate (DAP) other nitrogen materials will continue to diminish in importance.

4.15 The year 1984/85 was exceptionally good for fertilizer trade. World fertilizer nitrogen consumption increased by about 5% and trade increased by about 19% compared with the previous year. World ammonia trade also increased by about 11%. In the longer term, it is expected that there will be some significant change in the structure of international nitrogen fertilizer trade with the Developed Market Economies increasing their dependence on imports from both the Developing Market Economies and Eastern Europe. In the short term, however, the depressed agricultural situation, particularly in the Developed Market Economies will depress fertilizer demand in the next few years and slow down the longer term trend. Because of this, very little change is expected between 1985-1990 in world nitrogen fertilizer trading patterns.

4.16 There was a significant reduction in capacity in the USA following the plant closures in the early 1980s and in 1982 the USA became a net importer of nitrogen. With no new investment planned it looked as if the trend towards increased imports of ammonia would continue. The USA has for several years been a large importer of ammonia but most of this is processed and re-exported as diammonium phosphate. However, with a reduced demand in the USA in 1986 and 1987 as a result of the new farm bill and growing competition in the DAP export market, US ammonia import needs are now expected to decline for the next two or three years. The improved situation in the USA for ammonia producers with a reduced cost of gas will also slow down imports. After about 1989, however, unless the USA builds new ammonia capacity it will need to increase its imports and the most likely sources will be Canada and Trinidad.

4.17 A similar situation is forecast for Western Europe where in the longer term, nitrogen fertilizer exports will decline and net nitrogen imports will increase. Very little new capacity is planned and West European countries will become more dependent on Eastern Europe for their needs. The fact that European companies are contemplating investment in Trinidad could also increase exports from that source. Although urea consumption is making progress in Europe, ammonia nitrate is still the favored fertilizer to meet soil and crop needs so that increasing imports for the most part are likely to be as ammonia. In the short term, however, little change is expected to take place in Western European trade.

4.18 In Other Developed Countries and Oceania there will be a growing need for imports, particularly in Japan where nearly 30% of existing ammonia capacity will be closed under a rationalization plan. A new plant is being considered in Western Australia using offshore gas but this is not likely to come on-stream until the end of the 1980s.

4.19 Although new plants are projected to come on stream in Algeria and Nigeria by the end of the decade, Africa will remain a net importer. An export plant has been planned for Tanzania but this project is proceeding very slowly and is unlikely to come on-stream before 1990. If the Nigerian plant comes on-stream as planned in 1988, there could be a short-term export availability from this plant.

4.20 Latin America is a major net importer of nitrogen fertilizers due mainly to Brazil, Cuba and Mexico, but it is actually a net exporter of N due to the very large exports of ammonia from Trinidad, Mexico and Venezuela. It seems likely that the region as a whole will continue this overall trend to import more nitrogen fertilizers and export more ammonia. The gas-rich countries in the Caribbean area are well placed to compete for the growing import needs for ammonia in the USA.

4.21 With several new plants coming on-stream in the next few years, the Near East will strengthen its position as an exporter of urea and ammonia. New plants in Bahrain and Saudi Arabia will increase the ammonia export potential by about 0.5 million tons.

4.22 About 20 new ammonia/urea complexes are scheduled to come on stream in the next five years in the Far East mainly in India, and Indonesia. If the Indian plants come on stream as scheduled India will reach self-sufficiency by 1990 but it now looks as if some plants will be considerably delayed. Other plants in the region are also delayed and with an increasing growth rate of more than 5 tpy needs for nitrogen fertilizer imports will continue at more than 2 million tons of N throughout the remainder of the 1980s. This material will continue to come mainly from Eastern Europe and the Near East.

4.23 China dominates Centrally Planned Asia and as an importer seems likely to do so for some time to come. China's imports are likely to remain between 2-2.5 million tons N per year. It has been reported that in the longer term, China will be able to meet its own nitrogen fertilizer needs based on utilization of offshore gas at Hainan Island and could become a net exporter but this is not likely to happen before the mid-1990s.

4.24 The growth of Eastern European export potential has been spectacular. A massive construction program during the 1980s based on cheap natural gas has resulted in a surplus of more than 5 million tons of nitrogen supply capability, most of which is now exported in the form of ammonia and urea. Although this surplus will probably peak around 1986/87, the rest of the world will be absolutely dependent on imports from Eastern Europe throughout the remainder of the 1980s and beyond.

H. Countertrade in the Nitrogen Fertilizer Industry

4.25 Although countertrade has for many years represented a small part of international trade it has grown rapidly in the first half of the 1980s. With a shortage of money and credit a number of developing countries and centrally planned Eastern European have increased their use of countertrade as an essential means of doing business. Countertrade can involve various methods of trading such as:

(a) Barter - the simplest form of countertrade involving a direct exchange of goods or services of equal value. One of the largest and earliest countertrade deals in fertilizers involved the purchase of superphosphoric acid from Occidental Petroleum of the USA in exchange for ammonia, urea and potash from the USSR.

(b) Counterpurchase - the exporter agrees to purchase goods from the purchasing nation to a value related to the original exports.

(c) Compensation (or buy-back) - is similar to counterpurchase, except that the products exchanged are linked. For example, a fertilizer plant might be paid for in the form of product from that plant over an extended time period. Mitsui of Japan has an arrangement with the USSR similar to several others, where plant is provided in exchange for 100,000 tpy of urea and 100,000 tpy ammonia.

(d) Clearing accounts - the products to be traded with prices and trading duty are specified. At the end of the contract any imbalance is cleared.

4.26 Countertrade is believed to involve about 25% of nitrogen fertilizer trade at the present time and is growing, particularly in government to government deals. India has an agreement with the Soviet Union and the German Democratic Republic to purchase fertilizers in rupees which can be used by the supplying country to purchase Indian manufactured goods or raw materials. The USSR has also bilateral agreements with Pakistan and China for the provision of nitrogen fertilizers.

4.27 So long as Eastern Europe retains its dominance in the nitrogen fertilizer export market and until there is a major structural change in the balance of trade and debt relationships between the Developed Market Economies and the rest of the world countertrading seems likely to stay as a significant component of overall nitrogen fertilizer trade.

V. INVESTMENT AND PRODUCTION COSTS FOR NITROGEN FERTILIZERS

A. General

5.01 The two main factors in determining nitrogen fertilizer costs, the cost of feedstock and of investment, can vary significantly for different site locations and it is important to take these into account when projecting fertilizer costs and prices. Sometimes raw materials may be available cheaply, but this advantage can be offset by higher investment costs and lower operating rates if plants have to be built in remote locations and bear the cost of extensive infrastructure. The main purpose of this section has been to develop "cost envelopes" for ammonia and urea production which can be easily used to assess investment and total production costs for any situation. The comparisons have been made using "realization price." This is the price which would be required to justify new investment for an assumed location and for which the effect of the four main variables, (feedstock cost, utilization rate, capital charge and size of plant) have been evaluated. It is emphasized however, as used in this exercise, "realization price" is basically a total cost figure and does not necessarily reflect future fertilizer export prices, which will depend on many other market conditions.

5.02 It is difficult to measure the viability of a project in terms of a single parameter in view of the dynamic nature of the ranking of projects as the parameters change. A project which may appear to be more attractive than another project when based on a 15% IRR may be at a disadvantage when compared on a 10% IRR. Another aspect to be considered is the cash cost of production, which varies significantly from project to project. This is particularly important in an industry where sales prices fluctuate considerably and where, for a significant part of a project life, prices may fall to a level which would be insufficient to allow a company to meet its day-to-day costs.

5.03 In the following section, projects are compared on a 10% IRR and a 15% IRR and in certain cases over a wider range. Unless specifically stated plant utilization has been taken as 90% of design capacity. All financial data are in 1986 constant US dollars and an IRR model has been used, similar to that used in World Bank economic and financial analysis.

B. Investment Costs for Ammonia and Urea

5.04 An ammonia plant of 1,000 tpd has been assumed as the basis for comparing ammonia projects although data are also presented for a 1,350 tpd plant and a range of other plant sizes. Investment costs for urea are based on a 1,000 tpd ammonia plant and a 1,670 tpd urea plant. The investment costs are based on 1986 US dollars for plants expected to start-up in the period 1990-1995.

5.05 In some of the cost data that follow, it has been convenient to present realization price as a continuous function of investment cost, in other cases a scenario basis has been used to demonstrate more forcibly the impact of location on total production costs. The various scenarios considered are defined as follows.

(a) A developed site: In this case most of the supporting facilities will already exist, for example there will be roads, a port, railroad, a social infrastructure that will provide people to build and work in the plant, schools, hospitals, etc. Equipment can often be provided from local sources and maintained using local facilities.

(b) A developing site: In this case there will be some fertilizer and social infrastructure already existing which can usefully contribute towards the project but not as much as for case (a). Local specialized engineering services would not normally be available. It may be necessary to expand port or rail facilities to meet the needs of the project.

(c) A developing site - remote location: In this case there would be no supporting facilities of any sort available and all roads, ports, railways, civil works, amenities, etc., would have to be provided as part of the project cost. All equipment will have to be imported. Most of the labor to build and operate the plant will also have to be brought in from outside. There would be no supporting technical infrastructure.

5.06 Although costs in each category cover a relatively wide range, based on a large number of projects appraised and reviewed by the World Bank and also on information available on other projects, the following typical figures have been used as a basis for evaluation. In all cases, it is assumed that all power and steam are generated on the site using natural gas and that investment to do this is included in the offsites investments.

Ammonia Plant, 1,000 tpd (US$ million)

	Developed Site	Developing Site	Developing Site (Remote Location)
Battery Limits Plus Offsites	155	180	190
Infrastructure	-	40	80
Total Plant Costs	155	220	270

5.07 Working capital is not included as it varies with the price of gas, etc., and other production costs. Normally, however, it will be in the range of $6-10 million for a 1,000 tpd ammonia plant and this is automatically taken into account as part of the IRR calculations.

Ammonia/Urea Complex, 1,670 tpd Urea (US$ million)

	Developed Site	Developing Site	Developing Site (Remote Location)
Battery Limits Plus Offsites	230	260	280
Infrastructure	-	60	120
Total Plant Costs	230	320	400

5.08 In this case working capital is of the order of $10-15 million. When an additional plant is built on an existing site, there can be a significant savings in infrastructure costs and to some extent in offsites costs. In the estimates that have been made in Section VIII to assess the competitive nature of plants in different countries, additional plants on an existing site are assumed to cost 80% of the battery limits plus offsites costs.

5.09 It is felt that further comment on the investment costs given above is necessary. The last few years have seen a decline in real terms in the cost of building fertilizer plants and this is reflected in the detailed installed costs of several new projects which the Bank has had the opportunity to study and this information has been taken into account in preparing the investment data given in this review. However, some very low offers have been made during 1984 and 1985 to build new fertilizer plants. These low bids, mainly from Japanese companies, reflect the intense competition in the industry during a period when few new plants are being built. The data used in this review were examined critically to see whether allowance should be made for some further reduction in investment costs but it is believed that this is basically a short term situation and in any case is not typical of all new projects now being appraised. For example, the rapid appreciation of the Japanese Yen against the US Dollar in 1986 and the weakening of the US dollar against other currencies has meant that many of the low bids have had to be revised upwards.

5.10 It is also very important when comparing investment cost data to ensure that like is being compared with like and on the same basis. The investment cost data used in this report is regarded as the total cost of building and establishing the operation of a nitrogen fertilizer complex. A contingency figure has been added to the contractors' estimates to cover unforeseen expenses and a component is also added for preoperational expenses.

In order to ensure that all projects can be compared on a similar basis using a single energy source,the cost of the facilities to generate steam and electricity is also added. The total costs given in this report may therefore appear high when compared with engineering company costs who do not always include all these components.

C. Energy Requirements

5.11 The cost of energy both in the form of fuel and feedstock is by far the most important component in ammonia and urea production costs. This is clearly illustrated in table 15 which shows the average production cost of producing ammonia in the USA in 1985.

Table 15

Ammonia - US Production Costs 1985

	Cost $/metric ton	%
Natural Gas	95.14	75.96
Electricity	5.32	4.25
Other utilities	2.56	2.05
Operating Labor	2.33	1.86
Plant Overhead (within plant)	3.59	2.87
Maintenance Labor	1.27	1.01
Maintenance Overhead	1.00	0.80
Maintenance Material	2.67	2.13
Miscellaneous Supplies	0.81	0.65
Catalysts	0.97	0.77
Depreciation	7.23	5.77
Other	2.36	1.88
Total Production Cost	125.25	100.00

Source: The Fertilizer Institute Survey (TFI) of 4/29/86.

In the US in 1985 energy costs averaged about 80% of total production costs. Table 16, which is also the result of the TFI survey, shows how natural gas costs as a percentage of total ammonia production have increased since 1970. Due to the fall in oil prices in 1986, gas prices in the USA have also fallen and when the figures for 1986 are available, these may show some decline in energy as a percentage of total production costs.

Table 16

Natural Gas Costs for US Ammonia Production

Calendar Year	$/Million BTUs	$/Metric Ton of Ammonia	Percentage Of Total Production Cost
1985	2.65	95.14	76.0
1984	2.68	95.84	75.8
1983	2.34	91.24	72.0
1982	2.40	95.17	72.1
1981	2.33	93.51	72.6
1980	1.96	78.04	69.4
1979	1.62	65.03	66.8
1978	1.34	55.19	61.7
1977	1.14	46.27	57.2
1975	0.62	24.89	46.4
1973	0.36	15.30	45.6
1970	0.28	12.08	41.5

5.12 The TFI annual survey on energy usage in ammonia plants covering thirty nine US and Canadian companies and shown in Table 17 also provides a useful reference on actual energy consumptions.

Table 17

Average Energy Consumption in US Anhydrous Ammonia Production 1985 - Centrifugal Plants [a/]
Based on 10.65 million metric tons production

Energy Source	1000 BTU/metric ton of Ammonia			
	Feedstock Energy	Reformer Process Energy	Other Process Energy	Total
Natural Gas	24,556	13,265	3,039	40,860
Electricity		7	835	842
Imported Steam			(158)	(158)
Total	24,556	13,272	3,716	41,544

Source: Energy Use Survey for The Fertilizer Institute, 1985.

5.13 The TFI study is particularly useful as it relates to actual experience rather than engineering companies estimates of "battery limit" consumptions achieved under optimum conditions. Unfortunately, the latter figures are often used to calculate nitrogen fertilizer operating costs and give a misleading indication of the true cost of production, as actual energy costs usually exceed battery limit optimum consumption costs by a significant margin.

5.14 Under normal operating conditions, additional energy costs for offsites, handling, shutdowns and startups, maloperation, etc. may increase the guarantee figures by 15-20%.

5.15 About 0.58 tons of ammonia are required to produce one ton of urea and in practice about 9 million BTUs are required to produce one ton of prilled urea from ammonia. The TFI study shows the average energy usage in the US per ton of prilled urea including ammonia production is about 33 MMBTUs, based on an average ammonia energy consumption of 41.5 MMBTUs per ton.

5.16 Naturally a great deal of attention in recent years has focussed on reducing the energy requirements of ammonia plants. The overall effect of process development on energy consumption is shown in Table 18. Energy consumption per unit of product has fallen from about 75 MMBTU/metric ton of ammonia in the early 1940s to about 37 MMBTU/metric ton for the average plant built in the last few years. Most of the major ammonia process designers now offer low energy designs that are likely to result in total energy consumptions in practice around 32 MMBTU/metric ton of ammonia. A few plants based on such designs have recently commenced operation and others are under construction. Such designs are not the result of major technical "break-throughs" but more the integration of a number of energy conservation features some of which have already been used in ammonia plant design. Further major decreases in energy consumption may no longer be achievable since modern energy consumption for present process schemes are close to the theoretical and practical levels.

5.17 Table 18 gives a broad classification of world ammonia capacity and energy use.

Table 18

World Ammonia Capacity and Energy Consumption

Period Capacity Commissioned	Capacity Million Tons Ammonia	Energy Consumption MMBTU/Ton Ammonia
Prior to 1955	2.42	56
1955-1960	1.69	49
1960-1962	3.18	45
1963-1965	6.79	44
1965-1975	53.49	39
1975-1982	46.93	37
	114.50 a/	

a/ Total capacity includes 5.15 million mt/year for which the start-up date is unknown.

Source: Roger Heath and others, The Potential for Energy Efficiency in the Fertilizer Industry, World Bank Technical Paper 35, Washington, DC, 1985.

In a fully integrated energy plant for the production of ammonia and urea, gas is used to supply total energy such as electricity, steam and fuel as well as feedstock. Based on a new plant of modern design operating at 90% efficiency, it is estimated that the total energy consumption per ton of ammonia would be 32 MMBTU and for one ton of bagged urea, 28 MMBTU.

D. Realization Prices for Nitrogen Fertilizers

5.18 The bases for calculating realization prices are given below.

Total Energy Requirements. Ammonia - 32 MMBTU/ton

Urea - 28 MMBTU/ton bagged product

Other Variable Costs. As indicated in Table 15 other variable costs for ammonia are relatively small. These include catalysts, chemicals, cooling and boiler water, etc., and based on TFI and other specific project information a cost of $5.0/ton of ammonia has been assumed. The figure of $18/ton assumed for urea includes $13/ton for bags.

Fixed Costs. Labor and supervision costs have been estimated at $2.5 million/year for a 1,000 tpd ammonia plant and $3.7 million/year for the ammonia/urea complex. General administration and sales have been taken as $2.0 million/year for the ammonia plant and $2.8 million/year for the ammonia/urea complex. Capital related costs to cover maintenance, insurance and local taxes have been taken as 4% of plant investment cost and 2% of the infrastructure cost.

The IRR Model. The following parameters are assumed for both ammonia and urea.

Operational Life of Plant	15 years
% Investment phase in per year	20:40:35:5
% Production phase in per year starting in 4th year	70:90:100
% Phase in of working capital starting in 4th year	70:20:10
Salvage value of plant	10% of plant investment plus 100% of working capital
Working capital	Calculated by model depending on inputs

5.19 The information given in Annexes 15 and 16 covers most of the operating parameters that would be used to assess and compare the viability of ammonia and urea projects. The ideal situation for any project is the one in which cheap gas is available to a plant on a developed site, which can be operated at high efficiency to serve a local market. Few new projects nowadays will meet this situation although many plants in the USA did so during the 1960s and 1970s. Today, with changing trade patterns, the project economics will often involve a comparison of an export-based project on a developing site using low cost gas competing in the domestic market of an overseas country where gas is more expensive but the investment cost is much lower. The deciding factor in most cases will be the cost of transportation of the product to the market place. This particular issue is discussed more under the heading of fertilizer realization prices and the discussion below is related to ex-factory realization prices.

1. Ammonia Realization Prices

Details of ammonia realization prices for a range of gas costs, utilization and other factors are given in Annex 15 and Figures 5, 6, 7 and 8.

Developed Sites. These locations are most likely to apply in the USA and Western Europe where natural gas is available and competes with oil as an energy source. Based on a projected oil price of $18/BBL a gas price of $2.5/MMBTU has been assumed. (The basis for relating fuel costs is given in Figure 20.) The investment cost at this type of location is likely to vary from about $100 million for a 1,000 tpd ammonia plant on an existing site to about $150 million for a greenfield location. On this basis the estimated realization prices for ammonia would be approximately as follows.

Site	10% IRR	15% IRR
Greenfield	US$200/ton	US$230/ton
Existing	US$160/ton	US$180/ton

Developing Sites: On a developing site, particularly if it is a remote location, the capital charge will be the most important cost component. For a plant on a developing site to have the potential to compete in the export market it will be necessary to have gas available at less than about $1.0/MMBTU. If gas is available at $0.8/MMBTU, for investments on existing sites ranging from about $160 million up to $254 million for the greenfield plant, the ammonia realization prices will range between $145/ton-$200/ton for 10% IRR and from $175/ton to $245/ton for 15% IRR.

2. Urea Realization Prices

Details of urea realization prices are given in Annex 16 and Figures 9 and 10. Developed Site. For a range of investment costs of $180-250 million realization prices to give a 10% IRR realization prices are likely to range between $170/ton and $190/ton. For a 15% IRR the range would be about $190/ton to $220/ton. Developing Site. For a gas price of $0.8/MMBTU and a range of investment costs of $300-400 million, the corresponding range of realization prices is about $160-200/ton for a 10% IRR and $200-250/ton for a 15% IRR.

3. Effect of Utilization Rate

5.20 Utilization affects the capital charge cost component and so is much more important in situations where the investment costs are high. For example, on a developing site for ammonia production, a fall in utilization from 90% to 80% would increase the cost of capital recovery by about $25-30/ton. On a developed site, the corresponding figure is about half this. Although most projects are appraised on the basis of 90% utilization, world utilization of ammonia capacity is only 83% and in some developing countries, for example, the utilization is nearer to 70%.

4. Effect of Plant Size

5.21 Because the relationship between investment and plant size is not linear, there is an advantage for larger-size plants. In the case of ammonia, the capital related charge per ton for a 1,350 tpd compared with a 1,000 tpd plant on a developed site is about $16/ton less. For a developing site, the comparable difference would be between $30-40 per ton. The penalties of reduced size apply mainly to a situation of a single plant on one site. Two smaller plants sharing the same offsites and infrastructure as a larger plant of the same overall capacity may not be at a significant disadvantage compared with investment and could in fact have more flexibility with regard to operation.

5. Realization Price to Justify New Investment

5.22 If a project is to be export-based, it is evident that natural gas should be available cheaply at $1.0/MMBTU or preferably less. Investment cost for an ammonia/urea complex should not normally exceed about $1,000 per ton of nitrogen annual investment and in order to achieve this, capacity of the plant should be as large as possible. Full technical support should be provided to ensure that the plant will operate at at least 90% utilization. Realization prices for ammonia and urea plants operating at 90% utilization with various gas prices are shown in Figures 5-10.

E. Natural Gas Availability for Ammonia Production

5.23 Ammonia can be made from almost any source of hydrocarbon fuel or hydrogen but in recent years the process using the steam reforming of natural gas has become increasingly important. In 1982, as can be seen in Table 19, about 80% of ammonia production was from natural gas and the percentage is probably higher in 1986. Natural gas will certainly maintain and probably increase its position through the year 2000. Although consideration is being given to coal gasification as a major alternative method of production, the availability of cheap natural gas in many parts of the world and the relatively low investment costs necessary when processing this material will ensure its economic advantage.

Figure 5:

AMMONIA PRODUCTION — 1000 T.P.D.
REALIZATION PRICE TO GIVE 10% IRR FOR VARIOUS GAS PRICES
AND INVESTMENT COSTS

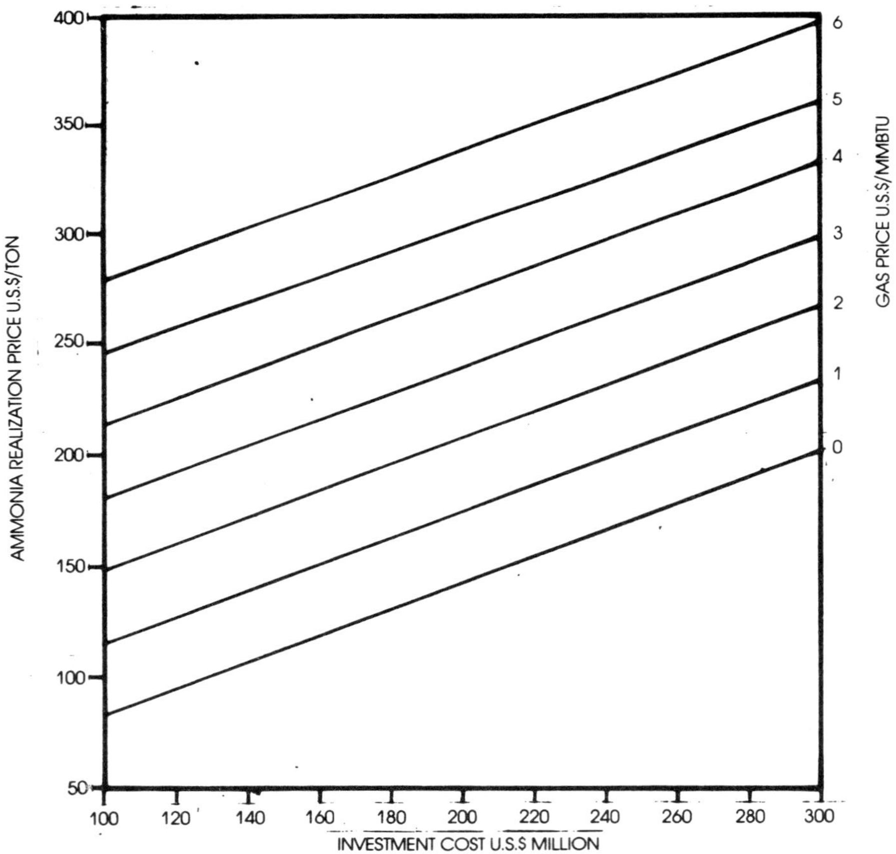

Figure 6:

AMMONIA PRODUCTION — 1000 T.P.D.
REALIZATION PRICE TO GIVE 15% IRR FOR VARIOUS GAS PRICES
AND INVESTMENT COSTS

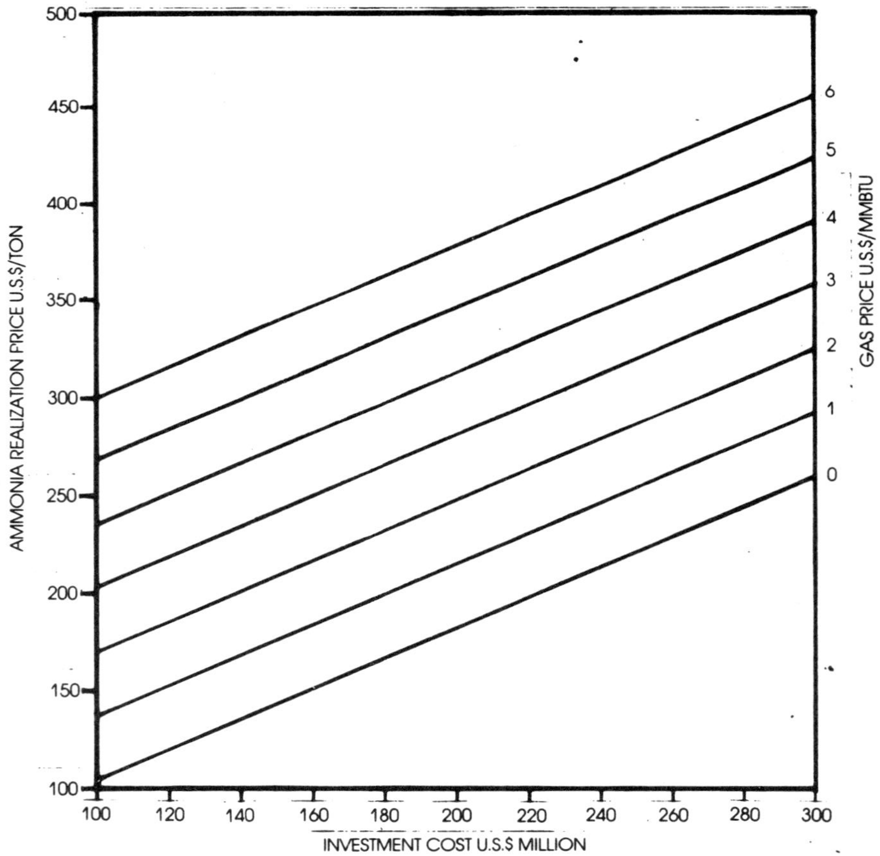

Figure 7:

AMMONIA PRODUCTION — 1350 T.P.D.
REALIZATION PRICE TO GIVE 10% IRR FOR VARIOUS GAS PRICES AND INVESTMENT COSTS

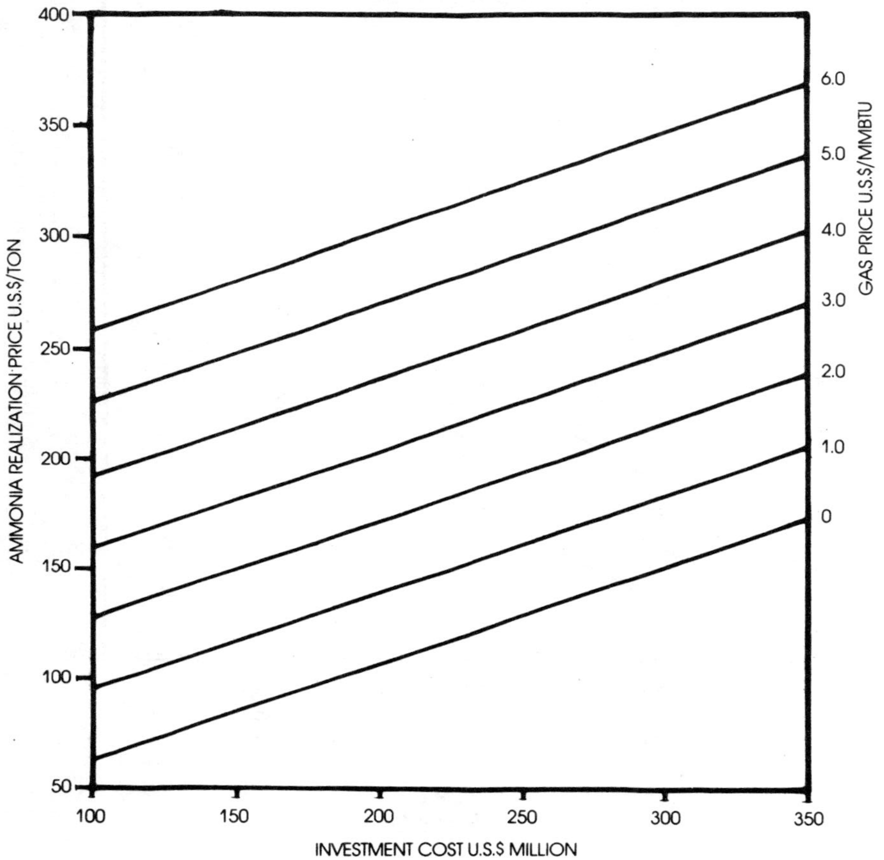

Figure 8:

AMMONIA PRODUCTION — 1350 T.P.D.
REALIZATION PRICE TO GIVE 15% IRR FOR VARIOUS GAS PRICES AND INVESTMENT COSTS

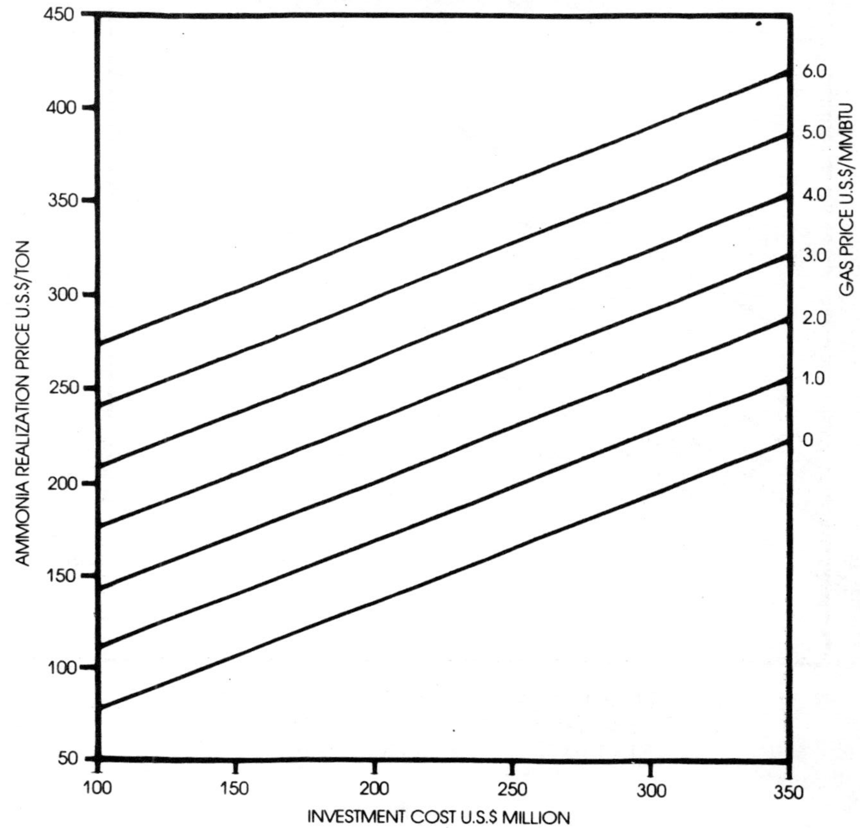

Figure 9:

UREA PRODUCTION — 1670 T.P.D.
REALIZATION PRICE TO GIVE 10% IRR FOR
VARIOUS GAS PRICES AND INVESTMENT COSTS

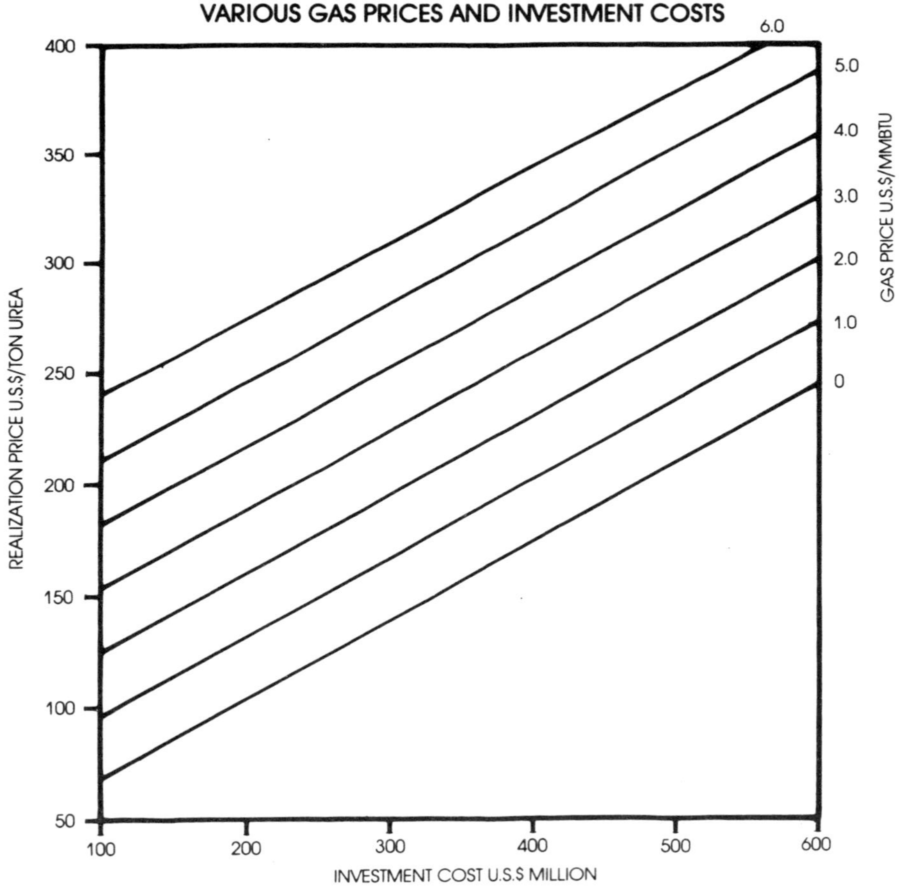

Figure 10:

UREA PRODUCTION — 1670 T.P.D.
REALIZATION PRICE TO GIVE 15% IRR FOR
VARIOUS GAS PRICES AND INVESTMENT COSTS

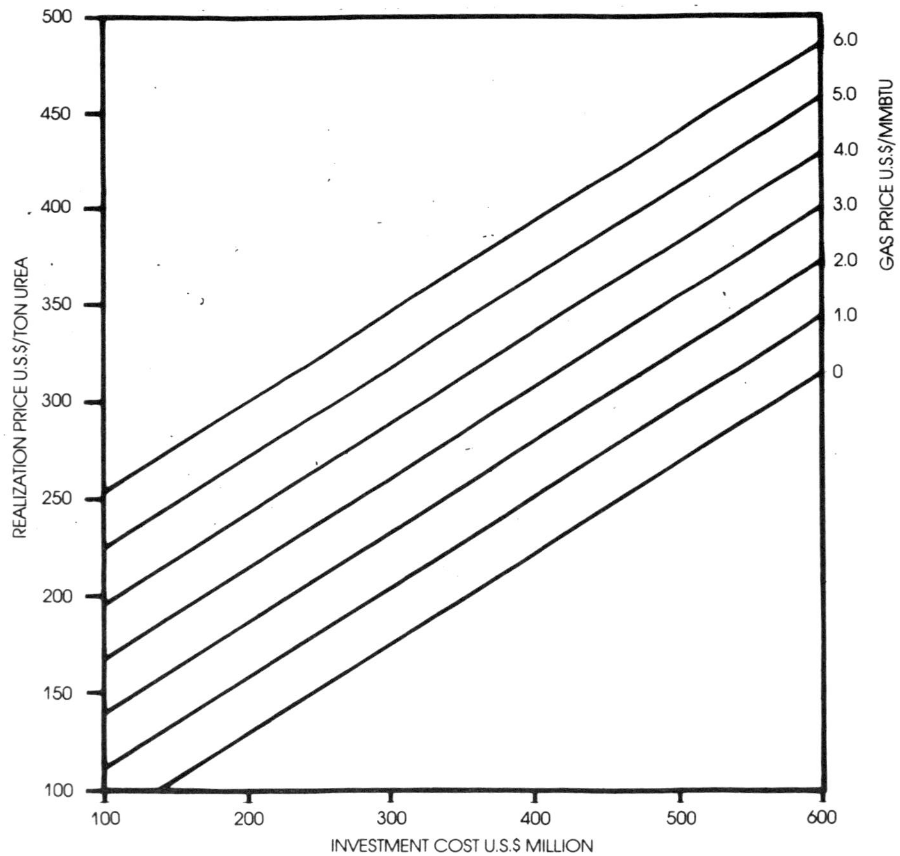

Table 19

Make-up of World Ammonia Production Capacity 1982

	Ammonia Production Capacity Million tpy	% of Total
Natural Gas Reforming	94.5	79
Naphtha Reforming	10.9	9
Refinery Gas Reforming	1.9	2
Fuel Oil Partial Oxidation	4.1	3
Coal Based Processes	6.3	5
Electrolytic Processes	1.0	1
Others	1.0	1
Total	119.7	100

Source: Heath and others. The Potential for Energy Efficiency in the Fertilizer Industry.

Details of natural gas production and reserves in the world are given in Annex 14 and some general comments on the major change taking place in the industry are discussed below. Specific comments regarding the potential of natural gas as a feed material for the nitrogen fertilizer industry are included in Annex 20 on country profiles. A summary of the production and disposition of natural gas is shown in Table 20.

Table 20

Natural Gas Production and Reserves

(Billion M^3)

Region	Reserves (1984)	Production (1983)
North America	8,258	521.54
Latin America	5,431	73.18
Western Europe	5,340	178.02
Africa	5,832	45.90
Middle East	22,436	39.55
Far East	6,390	91.34
Eastern Europe	36,638	598.21
Total	90,325	1,547.74

Source: Petroleum Economist, August 1984, Vol. II, Number 8.

5.24 In 1983 about 100 billion M^3 or about 6% of natural gas production was flared - equivalent to feed about 200 large ammonia plants or almost equal to the current world production of ammonia. The main deposits of natural gas are in the USSR, the Middle East and North America. Significant quantities of gas are being flared in the USSR, the Middle East and in several countries in Africa such as Algeria and Nigeria although the reduction in oil production in the last few years has reduced the availability of flared gas significantly. About 45% of the world's flaring takes place in the Middle East.

5.25 There are probably more than fifty countries that have the reserves available to develop an ammonia industry based on natural gas and about half of these are already producing ammonia. One of the most important factors determining the feasibility of ammonia production in any country will be the economic (opportunity) value of natural gas. This can vary significantly from location to location depending on the size of the resource and the alternative (opportunity) uses of the gas. If the gas can be used for oil substitution then the economic value of the gas should be linked to the value of the fuel oil. However, in many countries, particularly energy rich developing countries, oil substitution is not available and the value of the gas is determined by other alternatives in which the gas has a lower value. This issue is discussed further in Section VIII C.

VI. WORLD NITROGEN FERTILIZER PLANT CAPACITIES AND INVESTMENT NEEDS, 1990/91-1995/96

6.01 Based on current projections for increased nitrogen demand and anticipated needs to replace existing plants, some rough projections have been made of new capacity and investment needs for the period 1990/91-1995/96. In estimating demand, the increasing needs of nitrogen for the industrial market have also been considered.

6.02 World nitrogen fertilizer consumption is expected to increase by about 10.3 million tons between 1990/91 and 1995/96. Almost all of this nitrogen will be produced in the form of ammonia and then converted to either urea or other nitrogen fertilizers. In calculating the new capacity needed to meet increasing demand, allowances must also be made for distribution and process losses and also for the fact that plants usually work at some level below their design capacity. Distribution and process losses are assumed to be 5% and an overall utilization of 90% has been assumed for new ammonia capacity.

6.03 World production of ammonia in 1984/85 was estimated to be about 87.6 million tons N and of this about 11.7% or about 10.3 million tons N was used for industrial purposes other than fertilizers. Growth of the industrial nitrogen market is expected to be less than for fertilizer nitrogen and an average annual growth rate of 2.0% is assumed between 1990/91-1995/96, equivalent to an increase in demand of 1.1 million tons (11.3-12.4 million tons).

6.04 Although for project appraisal purposes, ammonia and nitrogen fertilizer plants are assumed to have a life of about 15 years, in fact the actual average life is usually much longer. It is difficult to obtain statistical information on the average life of ammonia and downstream nitrogen product plants as there are still plants operating more than 30 years old and others which have closed with less than 20 years life. In this exercise, it is assumed that the average plant life will be 25 years. Although a plant can be maintained indefinitely by replacing old equipment, normally technical obsolescence would require a major refurbishing after about 25 years. This life, however, would depend to a large extent on the rate of technical development. Other factors which would also force a plant to close would be the termination of a gas contract or unfavorable economic or market conditions. In Annex 9, details are given of ammonia plant capacity commissioned each year since 1960 and which was still operating in 1980. This information excludes about 13.0 million tons of the small and medium size nitrogen plant capacity in China for which specific start-up dates are not available. Between 1980-85 it is estimated that about 3.0 million tons of capacity in the USA and about 1.8 million tons in West Europe have closed permanently. About 1.3 million tons have closed in Japan, but closures in other countries are small. Little information is available on closures in the USSR.

6.05 Table 21 summarizes the start-up of new plant on a regional basis in 5-year periods since 1960/64 and Table 22 indicates the relative age of plant operating today. This capacity has been adjusted to allow for recent closures in the USA and elsewhere.

Table 21

New Ammonia Plant Commissioned 1960-1985 ('000 tons N)

	1960/64	1965/69	1970/74	1975/79	1980/84
North America	1,141	5,586	2,679	5,684	1,718
Western Europe	827	3,997	5,384	2,496	1,636
Oceania & Other Dd.ME	173	1,396	963	528	511
Africa	-	-	75	-	375
Latin America	410	662	2,632	668	843
New East	196	272	853	1,320	1,713
Far East	54	1,211	1,566	3,789	3,834
East Europe	2,041	4,683	7,065	9,447	7,037
Socialist Asia	145	408	-	3,646	718
Total	4,987	18,215	21,217	27,578	18,385

Table 22

The Age of World Ammonia Capacity ('000 tons N)

Commissioning Period	Capacity Operating 1979/80	Capacity Closed since 1979/80	Capacity Operating 1985	Cumulative Operating Capacity	Cumulative Operating Capacity %	Years Age
Before 1960	2,824	1,385	1,439	1,439	1.3	>25
1960-64	4,987	455	4,532	5,971	5.5	>20
1965-69	18,215	2,290	15,925	21,896	20.3	>15
1970-74	21,217	1,929	19,288	41,184	38.1	>10
1975-79	27,578		27,578	68,762	63.6	>5
1980-84	18,385		18,385	87,147	80.7	>0
Chinese small and medium plants not included			12,970	100,117	92.7	NA
Capacity unaccounted for by age a/			7,917	108,034	100.0	NA
Total Operating Capacity 1984/85			108,034	108,034	100.0	

a/ Refers to plant where start-up dates are unknown and also to expansions in some existing plant since commissioning.

The results in Table 22 indicate that considerable investment may be needed in the near future to replace or refurbish old plant. By 1990 about 20% of current capacity will be older than 20 years. It is difficult to put a precise figure to the replacement capacity needs but based on the closure rate in the 5-year period, 1980-84, the rate will be at least 1-1.5 million tons per year through the rest of the decade but as the percentage of old plants increase, the replacement rate will also increase.

Table 23

World Nitrogen Plant Investment Requirements 1990/91-1995/96

New Demand		Nitrogen (N)
For Nitrogen Fertilizers	(million tons)	10.3
For Industrial Use	(million tons)	1.1
Subtotal	(million tons)	11.4
New Capacity Needed[a]/	(million tons)	13.4
New Plants[b]/		36
Estimated Investment	(US$ Billion)	6.7
Replacement Capacity [c]/	(million tons)	7.5
New or refurbished plants		20
Estimated Investment	(US$ billion)[d]/	3.7
Total New Plants		56
Total Investment	(US$ billion)[d]/	10.4

a/ Assuming 90% utilization and 5% losses.
b/ On the basis of 1,350 tpd ammonia plants operating for 330 days py.
c/ Assuming replacement of 1.5 million tons per year.
d/ With downstream fertilizer production facilities investment needs assumed on average to be $500 per ton of nitrogen capacity per year, including refurbished plants. This low figure assumes many plants are built on existing sites.

The estimates indicate that the world will need about 11 new or refurbished large ammonia plants per year to meet demand after 1990.

VII. NITROGEN FERTILIZER PRICES

A. Historical Price Data

7.01 Prices for ammonia and urea for the last 20 years are shown in Fig. 11 and in Annex 21. The ratio of these two prices are plotted in Fig. 12 which shows that both absolute and relative prices have fluctuated widely, indicating that the ammonia and urea markets to some extent are independent of each other.

7.02 No reliable export price information is available for ammonia and urea before about 1965 but since then, there appears to be a widely fluctuating cycle of 6-8 years amplitude. Following an increase of prices in the mid 1960s there was a period of oversupply when prices slumped. In 1974 there was a return to balance with a very sharp rise in prices which were also affected by a major increase in energy prices at that time and fears of a major fertilizer shortage following the discussons of the World Food Conference in 1974. Prices at the end of 1974 rose to unprecedented heights as many developing countries, fearful of a forthcoming shortage of fertilizers, bought more than their immediate needs. This shortage was, however, short lived, since the rapid increases in prices slowed down demand and prices fell in the following years, particularly as new capacity which had been constructed during the shortage period came on-stream.

7.03 In 1979, a surge in nitrogen demand both in fertilizers and non-fertilizers brought about a more balanced situation and increased prices. With more limited capacity available, urea prices were considerably higher than for ammonia and peaked at $240 fob (bagged) in 1980. During the 1970s there had been considerable investment in ammonia plants in the USSR and when this new capacity came on stream at the end of the 1970s it helped constrain ammonia prices well below the level of urea prices.

7.04 The balanced situation continued through 1981 but a depressed world agricultural industry stagnated fertilizer demand at a time when increasing flows of ammonia and urea were becoming available from Eastern Europe. Between 1981 and 1983 urea prices almost halved. In order to sell their increasing surplus of ammonia and urea to obtain sorely needed hard currency, East Europe continued to displace other producers from the market by reducing export prices below the cash costs of many US operating companies and between 1982-1983, about 5 million tons of ammonia capacity was idled.

7.05 In 1984, it appeared that the bottom of the price cycle had been reached and prices started to rise again in the face of a strong demand which increased nitrogen consumption by 10% in 1983/84. Demand increased again in 1984/85 by 5% and prices continued to improve. Since mid-1985, however, the prices of ammonia and urea have fallen steadily. Ammonia supply capability and nitrogen fertilizer production potential increased by nearly 2 million tons in 1985/86 and world demand is expected to decline slightly thus increasing the short-term oversupply situation. The depressed world agriculture situation, the impact of the US Farm Bill, and the fact that some part of increased demand in the last two years has gone into stocks are the main reasons for the recent check in demand. The urea market, in particular, is dominated by a few major purchasers such as China, India, USA, and the failure of China to purchase significantly since its major imports of 1984, has had a depressing impact on the market since then. During the second half of 1985 and through 1986 ammonia prices fob Western Europe have declined from \$150/ton to less than \$100/ton. In the same period the fob price of bagged urea has fallen from \$180/ton to \$105/ton. In East Europe the fob price of bulk urea fell to \$65/ton in July 1986. With an increasing surplus of nitrogen over the next two or three years, there seems little hope for much improvement in nitrogen fertilizers prices during that period.

B. Some Considerations of Nitrogen Fertiizer Pricing

7.06 A recent paper by the World Bank describes some of the factors affecting fertilizer pricing principles and the difficulties of making fertilizer price projections due to market imperfections. In general the major factors are:

(1) The limited numbers of buyers and sellers

(2) Limited entry or exit of producers to the industry.

(3) Increasing internal and external returns to scale in production.

(4) Long time lags in adjustment to market equilibrium.

(5) Distortions in input prices and limited mobility of reserves.

(6) Imperfect knowledge of market conditions.

(7) Effect of countertrade and barter deals.

(8) Exchange rates.

With regard to the first factor in this list, the world nitrogen industry is now dominated by two countries in Eastern Europe - the USSR and Romania, which together comprise nearly 40% of exports in nitrogen materials (ammonia plus nitrogen fertilizers). Likewise, demand for inputs is strongly influenced by four major countries, China, India, Iran and USA. Apart from the USA, import decisions in these countries are highly centralized and are sometimes influenced by political as well as economic considerations. More than 70% of the world's nitrogen fertilizer industry is now government controlled and is protected in many countries (for strategic and social reasons) from free international market competition.

7.07 The very large investment costs required for new nitrogen fertilizer production complexes inhibit easy entry to the industry but once in, because of the high capital charges costs, plants are usually operated at high output to maximize contribution. As mentioned earlier there appears to be a price cycle of about 6-8 years. This may be due to the fact that it takes 3-4 years to implement a new project which is about half frequency period of the price cycle.

7.08 Within this cycle, and where free market conditions do exist, (essentially in the USA) the prices of ammonia and urea tend to be determined by the following factors:

(a) The relative balance between potential supply and demand.

(b) In a surplus situation, the cash cost of production in those plants most vulnerable to closure (excluding these plants protected by Government action in the form of subsidies or penalties for closure).

(c) In a tight situation the price of product of which demand is restricted by affordability.

The USA has now become the balancing market for nitrogen products particularly ammonia and this was clearly demonstrated during the early 1980s when nearly 5 million tons of ammonia capacity were idled in the face of severe competition from imports. Western Europe and Japan also suffered, but to a much lesser degree because of industry protection by their governments. The sensitivity of fertilizer demand relative to the state of the US agricultural industry and government farming policies also makes the US fertilizer industry much more responsive to free market conditions than is experienced elsewhere in the world.

7.09 The problem of forecasting fertilizer prices is also complicated by the fact that export material applies to only about 20% of total production. Domestic prices for a number of reasons, including government control, often differ significantly from export prices.

7.10 Another factor mentioned before which affects fertilizer pricing is the imperfect knowledge of market conditions. This issue is accentuated at the present time when, although the USSR dominates the world nitrogen market, very little is known by the rest of the world of the economic aspects of manufacture or plans for future investment in the Soviet Union.

C. <u>The Relationship between Ammonia and Urea Prices</u>

7.11 The relationship between urea and ammonia export prices fob West Europe is shown in Fig. 12. Although there has been considerable variation in the ratio over the years generally this ratio has been declining. The reason for this general decline is not evident but undoubtedly the relative increase in growth of urea capacity has made urea much more available over the years and reduced its cost. Another factor is that the relative investment costs for urea plants compared with ammonia plants has declined over the last two decades from about 50% to 30%.

7.12 The theoretical relative costs of producing ammonia and urea on the same basis have been derived from the results in Annex 16. Covering the different site locations it was found that the realization price ratio for bagged urea compared with ammonia varied between about 0.97 to 0.93. In a well balanced market for both ammonia and urea this ratio would probably hold. In a tighter market situation where urea capacity is probably the constraining fact the price ratio is likely to rise.

7.13 On the other hand in a weak market where there is both excess ammonia and urea capacity a producer may have to decide how to optimize his overall combination according to prevailing export prices. In this situation, on the basis of 0.6 tons of ammonia per ton of urea and assuming cash costs of about $25 for producing one ton of bagged urea from 0.6 ton of ammonia it might be expected that producers would be prepared to manufacture urea for a price which could

Figure 11:
AMMONIA AND UREA PRICES 1968-1985 — Current US$

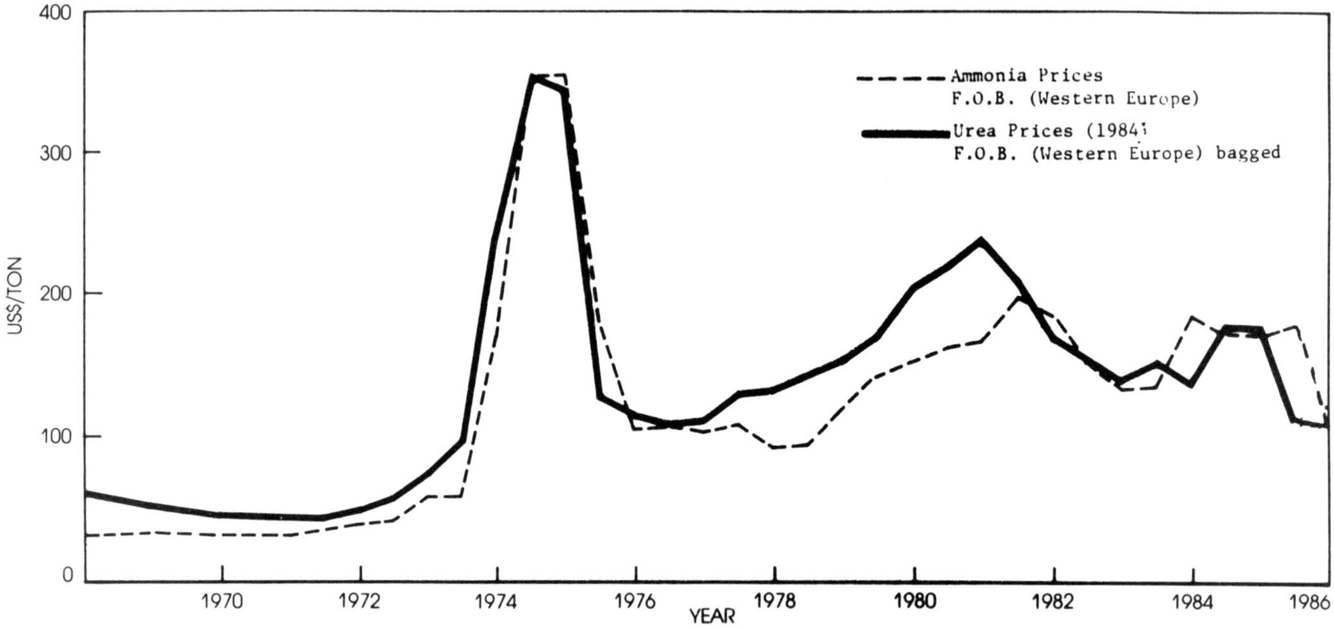

Figure 12:
RATIO OF UREA: AMMONIA F.O.B. EXPORT PRICES WEST EUROPE

World Bank—30178:4

fall to 0.6 x price for ammonia + $25 per ton of bagged urea. In this situation the ratio of urea: ammonia prices is likely to fall.

D. The Relationship between Ammonia Capacity and Production Costs

7.14 In order to establish a better basis for understanding the current cost structure of world nitrogen production, some estimates have been made of the relationship between cost of production (both cash cost and cash cost plus capital charge) against world capacity for ammonia.

The following parameters were used in the calculations.

	Energy Consumption Age of Plant-Years	Energy Consumption MMBTU/Ton of NH_3
1	0-5	36
2	5-10	40
3	10-15	44
4	>15	46

Energy Costs. Where available, specific energy costs have been used. In the USA, gas prices were apportioned to capacity according to the 1985 TFI survey. It has been assumed that in the energy-rich countries of the Near East, USSR and North Africa existing gas prices are of the order of $0.5/MMBTU.

Investment Costs

Developed Site	US$300 - 500/ton N capacity/year
Developing Site	US$500 - 700/ton N capacity/year
Developing Site (remote location)	US$700 - 900/ton N capacity/year

Cash Costs. These include variable costs plus certain non-capital related fixed costs such as labor.

Capital Related Costs. These have been related to the age of the plant to cover depreciation,maintenance, insurance and interest (10%). It is assumed that plants are depreciated over 15 years but for plants more than this age an additional maintenance charge has been included. The ammonia capacity is based on the Bank's 1986 ammonia plant list, totalling 108 million tons N, but excludes 16 million tons of Chinese capacity for which there is insufficient cost information. Plots of cash cost and total production costs against ammonia capacity are given in Figs. 13 and 14. This information shows a wide range of both cash costs and total costs for world ammonia production. It is important to appreciate that the sequence of cumulative capacity in the abscissae of Figures. 13 and 14 do not necessarily coincide. There is about 40 million tons of ammonia capacity with a cash cost of production less than about $50/ton. This acounts for most of the ammonia capacity in the energy rich countries where gas prices are known, or assumed to be very low. The cash costs for the remaining 50 million tons covered in the study increase steadily up to almost $230 ton. The high end of this range relates to some of the plants in Japan using either naphtha or imported LPG.[a/] Production costs including capital charges vary from about $85/ton up to $280/ton. The profile of the cost curves explains to some extent the ability of the export-oriented sector of the industry to operate at very low prices over a considerable period of time.

E. The Relationship between Nitrogen Fertilizer Prices and Ammonia Capacity Utilization

8.15 Exploratory data analysis was conducted to measure the degree of association between nitrogen fertilizer prices and ammonia capacity utilization. Ammonia capacity data were based on the plant list of the World Bank/UNIDO/FAO Industry Fertilizer Working Group for 1985. Nitrogen production was based on FAO historical data for nitrogen fertilizers plus the Working Group estimates of ammonia used for industrial use. No allowances were made for ammonia losses during processing to down-stream nitrogen products and this is

---

a/ Based on energy prices prevailing in 1985.

assumed to be a function of ammonia production (based on past experience - about 6%). Nitrogen capacity utilization was smoothed to eradicate violent yearly fluctuation and plotted on a lagged basis, against urea and ammonia export prices. In view of the exceptional factors which led to massive price hikes in 1974 and 1975 these data points were not included in the analysis.

7.16 A correlation coefficient in excess of 85% was found to exist between urea prices and nitrogen capacity utilization up to the early 1980s. In the case of ammonia the correlation was much weaker and of the order of only 60%. Non-linear transformed and lagged relationships were tried but the coefficient still remained low.

7.17 A linear regression model was used to attempt to predict urea prices based on projected plant capacity utilization to meet future demand. As would be expected by looking at Fig. 15 rather high values of future urea prices were obtained although the standard errors were also high. Non-linear relationships were also tried but the plot of predicted versus actual values showed very large residuals. A weighted least squares procedure which heavily weighted the recent observations and discounted the distant past was also utilized but the R^2 turned out to be unsatisfactory for predicting future prices. It is believed that two major reasons for the difficulty in predicting future prices using statistical techniques with historical data is the changing structure of the market with increasing dominance by Eastern Europe, and the changing value of the dollar with other currencies.

7.18 This latter part is demonstrated in Figs. 17 and 18 which show the relative historical prices of urea in US dollars and Indian rupees. Between 1964-1980 the relative value of prices was reasonably constant but in the last few years there has been a major change in the relative values. In terms of US dollars urea prices have fallen sharply in the last few years but in terms of many other currencies the reverse is often true.

7.19 Perhaps the most important reason for the lack of correlation between capacity utilization and nitrogen fertilizer prices is due to a major change in the structure of the industry that has taken place since about 1979. In this period about 25 million tons of new ammonia capacity has been installed mainly in Eastern Europe or the Developing Market Economies and most of this capacity is based on low cost natural gas. To a large extent this explains the large percentage of the world's nitrogen industry now capable of operating at very low cash costs. Although there is a significant surplus of nitrogen supply capability, in relative terms, it is much smaller than the case of the other two major nutrients, phosphate and potash. For example, there is currently a surplus supply capability for nitrogen of about 4% whereas in the case of phosphate and potash the surplus is about 10%. Nevertheless, with exports representing only about 20% of total nitrogen fertilizer production, whereas capacity with low cash cost, including most of the export-based plants represents nearly 40% of the industry, prices can remain very low even at a time when there is a relatively small surplus of supply. This is particularly so when the market is dominated by a few major importers in the position to force prices downwards, and where the major exporting countries from Eastern Europe seem determined to obtain markets at any price in their search for overseas earnings. It is not too surprising therefore that the response function between surplus and prices is no longer responsive as it was during the 1970s. This situation seems likely to continue for the next two or three years but when a balanced situation arises, as anticipated at the end of the 1980s, prices could then rise very sharply.

Figure 13:
AMMONIA CASH COSTS AGAINST WORLD N CAPACITY

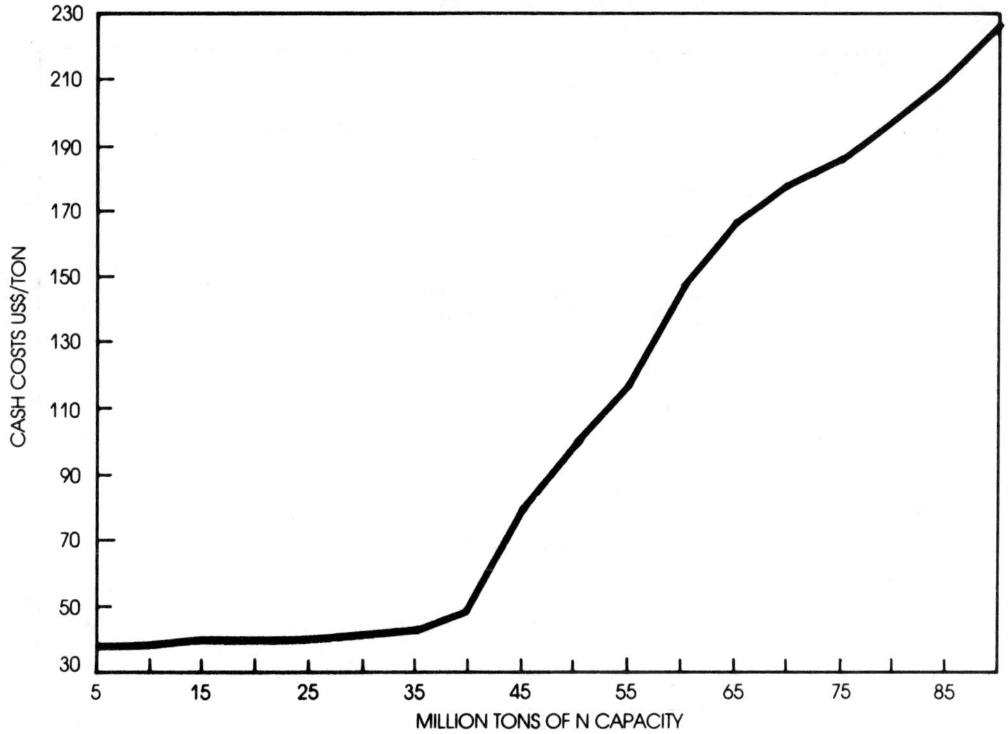

Figure 14:
AMMONIA PRODUCTION COSTS AGAINST WORLD N CAPACITY

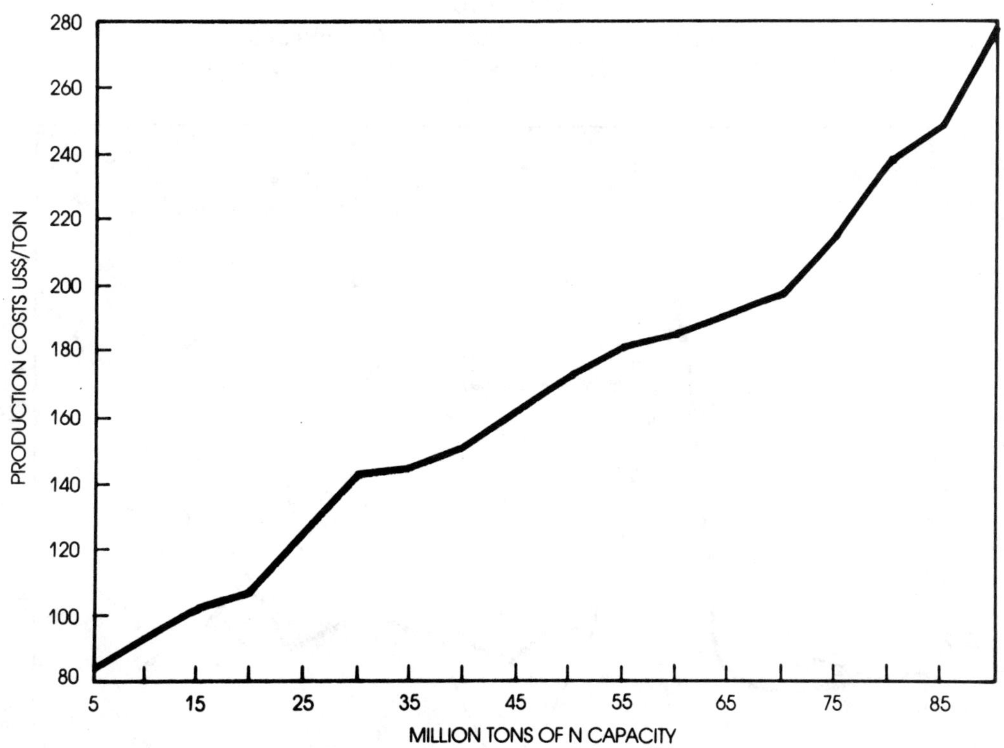

World Bank—30178:5

Figure 15:
VARIATION OF FERTILIZER MATERIAL PRICES WITH AMMONIA CAPACITY UTILIZATION

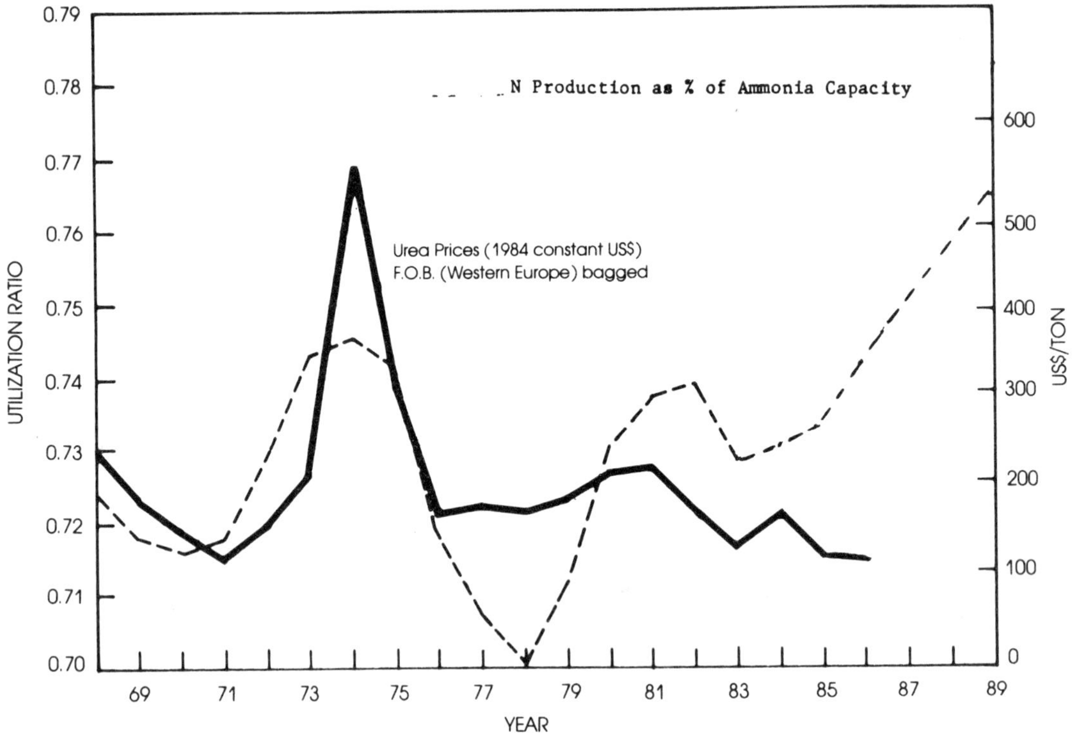

Figure 16:
VARIATION OF FERTILIZER MATERIAL PRICES WITH AMMONIA CAPACITY UTILIZATION

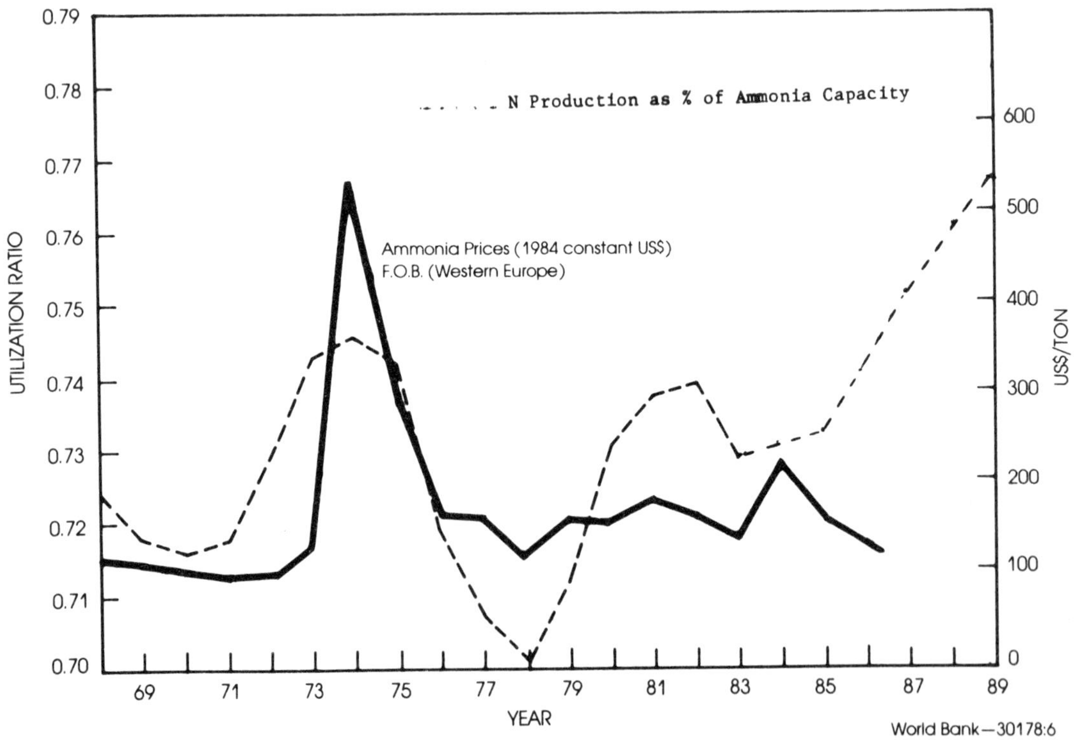

World Bank—30178:6

Figure 17:

UREA PRICES IN INDIAN RUPEES

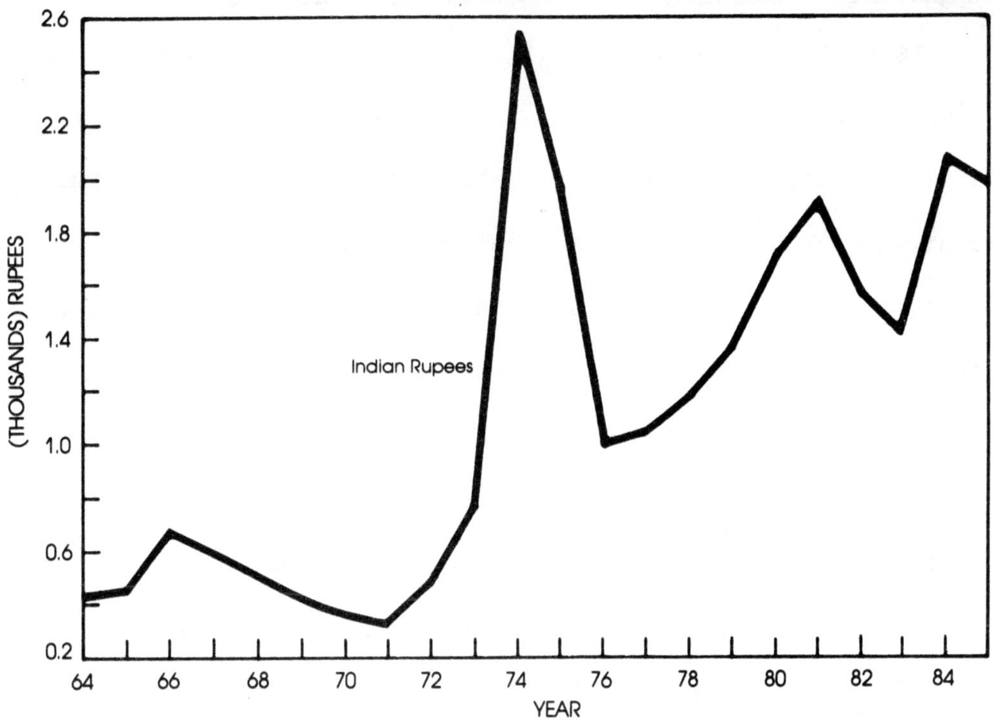

Figure 18:

UREA PRICES IN US DOLLARS

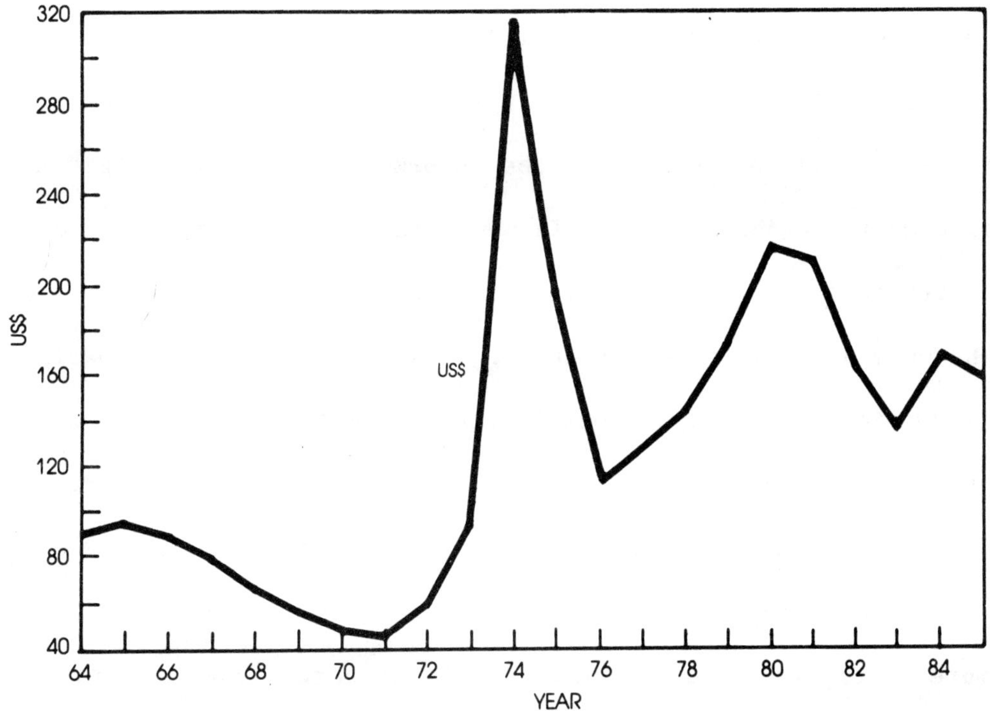

World Bank—30178:7

F. Investment and Realization Price Data for Potential Nitrogen Projects

7.20 It is forecast that excess nitrogen capacity will continue to increase for a year or two and then the surplus will diminish quite quickly to reach a balanced situation about 1988/89. Thereafter about 11 new or refurbished plants will be required each year to meet increasing demand and some replacement of very old plants. In order to assess the competitive position of potential new projects in the nitrogen fertilizer market, a matrix has been constructed to compare the fob and cif realization prices for bagged urea and ammonia for potential new projects. Realization prices have been estimated to give a 10% and 15% IRR for various gas prices for which it is assumed contracts could be obtained in 1986. Investment has been considered on the basis of a completely new site and also on an existing site. In the latter case it is assumed that the investment cost would be 80% of the battery limits and offsites investment cost that would prevail for a similar plant on a new site. Cash costs are also stated and a range of landed costs to give an 10% and 15% IRR for the various producers in the major future markets of North America, West Europe, S.E. Asia and the Far East have been derived. The information for urea is given in Annex 17 and for ammonia in Annex 18. It is emphasized that the realization prices stated are not forecasts of future market prices for ammonia and urea and these will depend on many other market factors. Also in making the comparisons of the relative economics of different location, an attempt has been made to assess investment on an equitable basis but taking into account the different location factors. It is appreciated, however, that in practice in some cases, more favorable investment costs and financing may be obtained which would make the project more attractive than indicated in the comparisons given in Annexes 17 and 18.

7.21 A summary of estimates of realization prices that would be required to attract new investment for plants that would come on stream in 1990-1995 is given in Table 24. Three locations, U.S. Gulf, West Europe and the Arabian Gulf have been used as a basis for the comparison. The realization price projections are based on the assumption that oil prices in 1986 US dollars will fall within the range of $17-20/BBL between 1990-1995. Although the comparisons are based on ammonia plants of 1,000 tpd, some new plants will have a larger capacity and in this case, the price of ammonia and urea will be reduced as indicated in Section V.

Table 24

Ammonia and Urea Realization Prices 1990/95

	Ammonia (1986 US$/ton)	
	Low Estimates	High Estimates
US Gulf	165	190
West Europe	175	200
Arabian Gulf	155	190
	Urea (1986 US$/ton - bagged)	
	Low Estimates	High Estimates
US Gulf	155	180
West Europe	165	190
Arabian Gulf	145	180

The figures given above are value judgements. The low estimates assume a plant on an existing site whereas the high estimates assume a greenfield site.

G. Ammonia and Urea Price Projections 1990-1995

7.22 Although realization prices are an indication of the future prices that would be needed to justify new investment, these should not be confused with actual fertilizer prices which will depend on many other factors. Price fluctuations in the past few years have clearly demonstrated the difficulties of making projections, particularly for ammonia and urea where cash costs of production in the industry range from about $50/ton up to $200/ton and the market is dominated by a few major suppliers and purchasers. World Bank fertilizer price projections are made by the Economics and Projections Department (EPD). The broad views given in this section are based on the Fertilizer Working Group's nitrogen supply and demand projections but there is complete agreement with the Bank official price projections made by EPD using their comprehensive agricultural/fertilizer model which is described in Section IX.

7.23 Nitrogen fertilizer prices are expected to remain depressed over the next two years during a period of surplus supply which will be sustained by reduced demand in North America and Western Europe caused by the depressed agricultural situation in these regions. After 1988, however, the surplus balances will diminish and a strong upwards price movement will commence and this will continue into the early 1990s. There is little new nitrogen fertilizer plant construction underway at the present time but it is anticipated that the prospects of higher prices will stimulate the construction of more new capacity at the end of the 1980s and this will come on stream in the early 1990s alleviating the pressure on prices as the mid 1990s are approached. The extent to which prices will rise and the period for which they will stay elevated will depend on the rate of new investment. It is, however, unlikely that the investment and construction rates will incrase to the levels achieved in the last investment cycle when more than 50 new plants were installed in the Centrally Planned Economies and there was also considerable investment in North America. The current five year plans in both the USSR and China have only limited investment plans for ammonia plants and the uncertainty over energy prices in many countries together with general economic problems are likely to have a dampening effect on new investment prospects.

VIII. THE EFFECT OF FUTURE ENERGY PRICES ON THE WORLD FERTILIZER INDUSTRY

A. General

8.01 The energy requirements and costs for ammonia and nitrogen fertilizer production have already been discussed in Section V B and this information is best illustrated in Fig. 19. For a new plant on a developed site where gas prices are likely to be relatively high and equivalent to other energy sources such as fuel oil naphtha, etc., total energy costs can be more than 60% of the total cost of production. In other cases where the gas is cheap such as a developing site, but where all infrastructure has to be provided, the cost of gas can be less than 10% of the total production cost. Obviously in this situation international oil prices can have a major impact on the overall production costs in Developed Market Economies and other countries where oil and gas prices are directly related and hence on the relative competitiveness of different projects. Before looking at this situation in more detail however, it is necessary to have a better understanding of the present situation with regard to energy pricing for nitrogen fertilizer production. Until 1970, before the major hike in oil prices the energy costs for nitrogen fertilizer production were probably closely related to world energy prices using crude oil price as a basis. At that time, almost all nitrogen capacity was in the Developed Market Economies, or the Centrally Planned Economies of Eastern Europe and many plants were based on coal, cokeoven gas, fuel oil, naphtha as well as natural gas. Generally the size of plants was small and plants served local markets. Trade in nitrogen fertilizer was relatively small and energy pricing at that time, although an important factor, was not an overriding one. The 1970s saw many

Figure 19

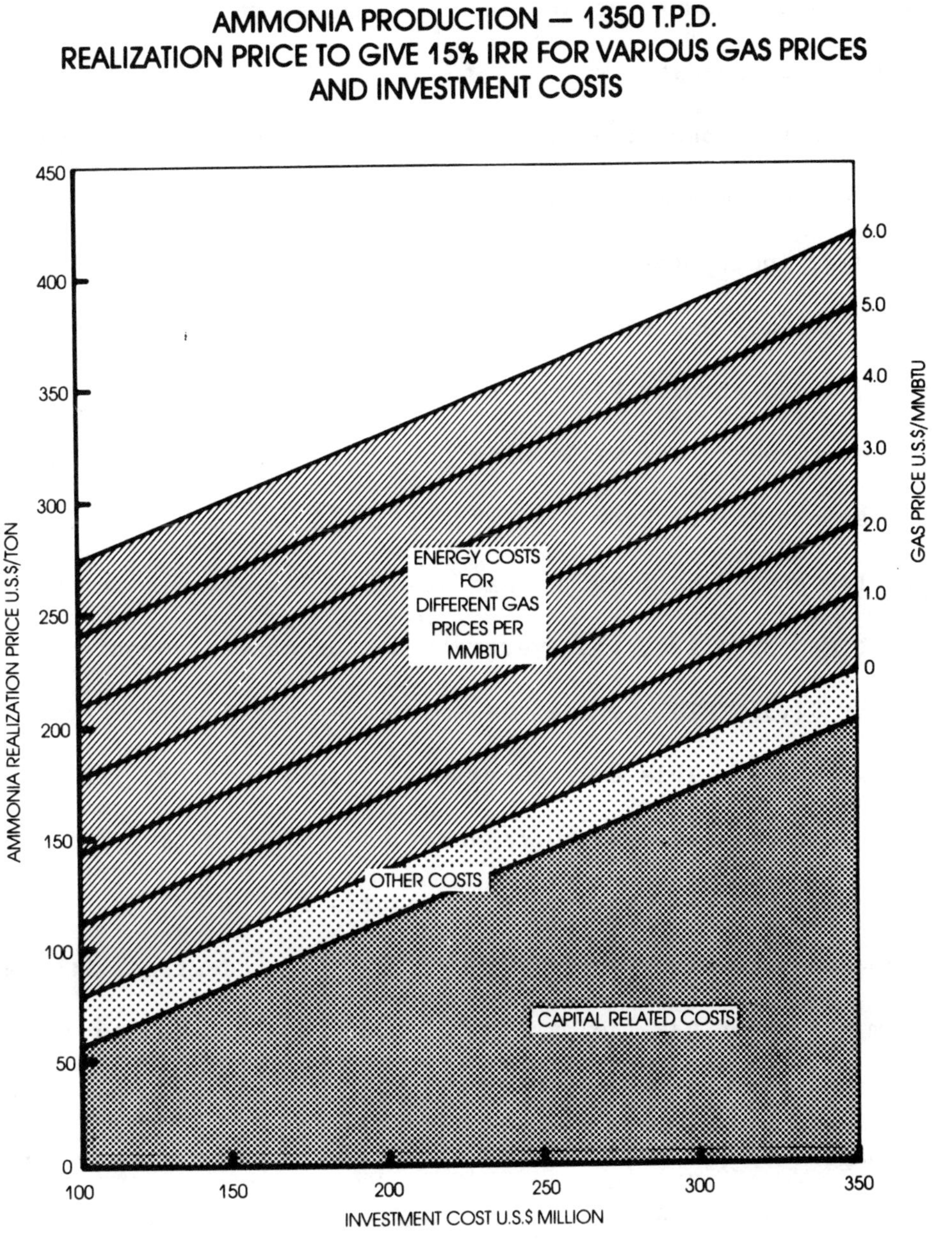

World Bank—30332:1

changes in the structure of the world nitrogen industry. The large rise in oil prices had a major impact on those plants based on fuel oil and naphtha, mainly in the developed countries and many were subsequently forced to close down or change to other feedstocks.

8.02 At the same time, in order to meet the rapidly increasing demand for nitrogen there was a major building program of new large ammonia plants based on natural gas, incorporating many new design features which improved energy efficiency and reduced production costs. In the Developing Market Economies and Centrally Planned Economies with large reserves of natural gas, the price of energy was well below the equivalent price of energy in oil. In some Developed Market Economies the price of gas as a feedstock for ammonia plants also stayed well below that of oil. For example in the USA with large available gas supplies, the average price of gas to ammonia plants in 1984 was still only $2.5/MMBTU whereas the equivalent price of energy in fuel oil was about $3.5/MMBTU. This average price of $2.5, however, included about 30% of low gas contracts (average $0.7/MMBTU) and 70% of market price contracts (average $3.0/MMBTU). In Europe, companies in the UK, Netherlands and Ireland had been encouraged to increase their ammonia production with relatively cheap gas from new major gas fields. Some countries, who had no cheap gas available however, were placed at a major disadvantage with regard to ammonia production and perhaps the most badly hit was Japan which found itself with a naphtha based industry in a very uncompetitive situation. Subsequently, exports of nitrogen products from Japan declined dramatically.

8.03 Even so, for most of the world's nitrogen industry, the price of gas has stayed well below the equivalent price of energy in oil. Based on the information collected for Fig. 13 it is estimated that in 1985 the average world price of gas to ammonia plants was still less than $2.5/MMBTU and probably about half of world capacity had a gas price of less than $1.0/MMBTU when the equivalent energy price of fuel oil was $3.5/MMBTU.

B. The Cost of Producing Ammonia with Different Feedstocks

8.04 Although natural gas is currently the most commonly used feedstock for ammonia production and is expected to remain so, there may be occasions, depending on relative energy pricing, when other feedstock may be competitive. A comparison of total production costs for different feedstocks is given in Figs. 20 and 21. The most important factors in determining production costs is the relative investment and energy costs which have been taken as follows for a 1,350 tpd ammonia plant.

	Ratio of Investment Costs	Total Energy MMBTU/Ton Ammonia
Natural Gas based Plant	1.00	32
Naphtha based Plant	1.15	35
Fuel Oil based Plant	1.60	37
Coal based Plant	2.00	45

Although there is much information available on the the investment and production costs of gas-based processes, information on comparable up-to-date processes for the other feedstocks is not so readily available. In this respect a relatively recent investigation by Tennessee Valley Authority a/ on comparison of ammonia production by natural gas and coal gasification was very useful. Because capital charge is such a major cost component the comparisons are given both on a 10% and 15% IRR basis in Figs. 20 and 21.

a/ D. A. Waitzman et al, "Progress Report on Ammonia Production from Coal by Utilization of Texaco Gasification Process," AICHE Meeting, November 1983.

Figure 20

AMMONIA REALIZATION PRICE VERSUS FEEDSTOCK COST
UTILIZATION 90%: IRR 15%

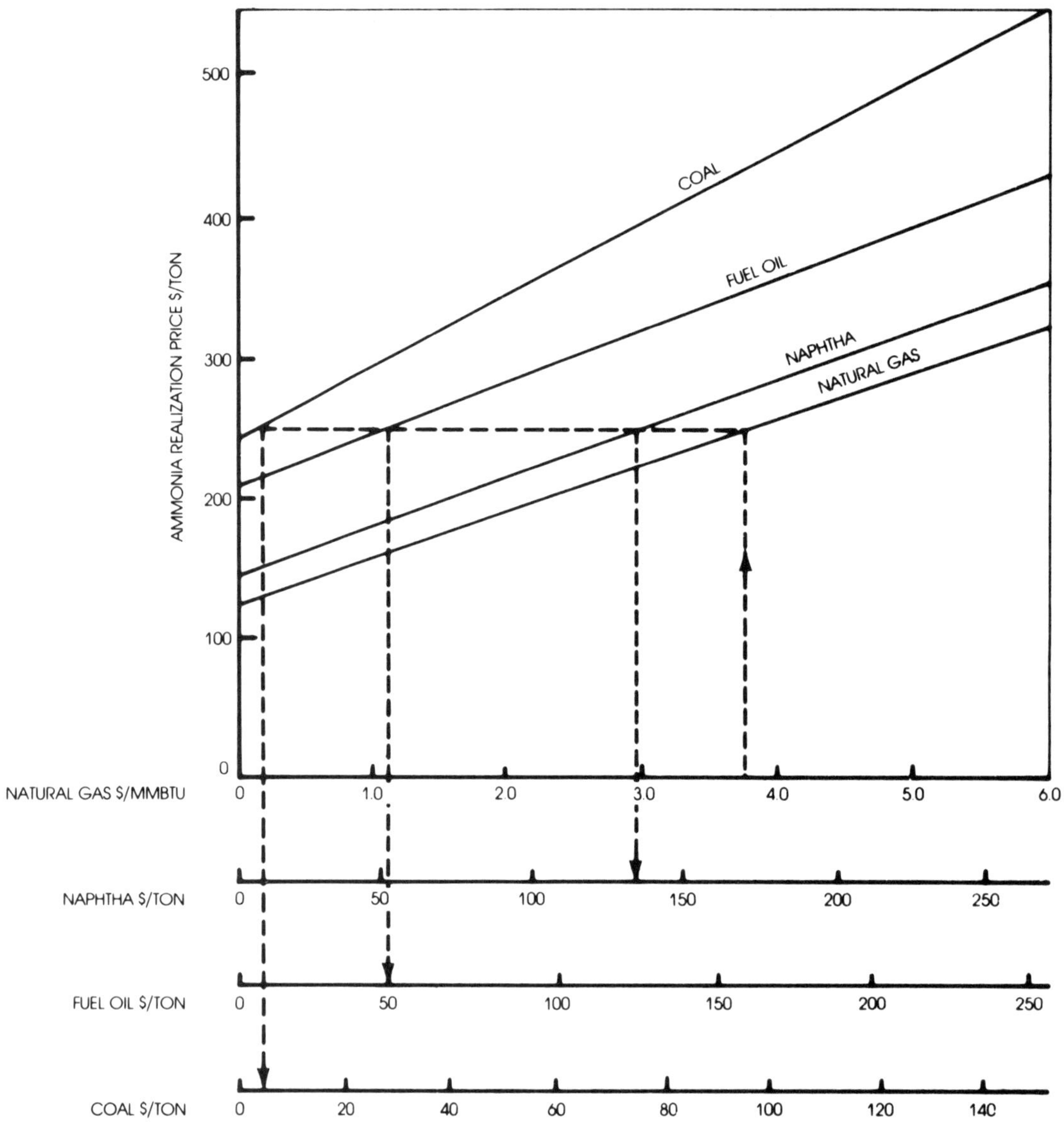

<u>Equivalent Energy Prices For Different Fuels</u>

World Bank – 30332:3

Natural Gas US$/MMBTU	Naphtha US$ Per Ton	Fuel Oil (No.6) US$ Per Ton	Coal US$ Per Ton	Crude Oil US$ Per bbl
1.0	44.8	41.9	25.1	7.1
2.0	89.6	83.8	50.2	14.2
3.0	134.4	125.7	75.3	21.4
4.0	179.2	167.6	100.4	28.5
5.0	224.0	209.5	125.5	35.6
6.0	268.8	251.4	150.6	42.6

The price of crude oil assumes that Fuel Oil No. 6 is 80% of the cost of crude oil on a weight basis and that there are 7.35 barrels of crude oil per ton of crude oil. The prices of natural gas, naphtha, fuel oil and coal are then equated on an energy basis.

Figure 21

AMMONIA REALIZATION PRICE VERSUS FEEDSTOCK COST
UTILIZATION 90%: IRR 10%

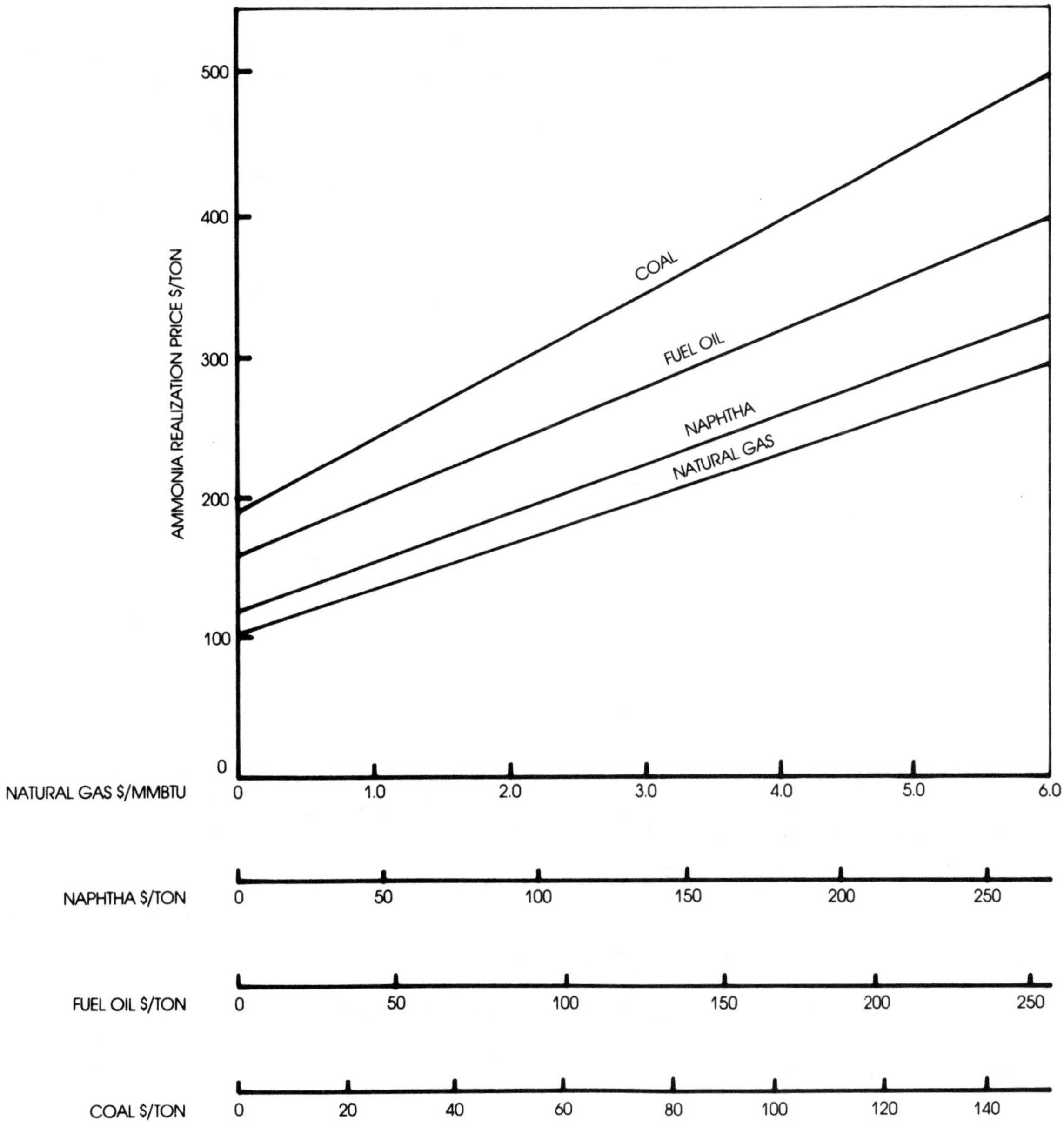

World Bank—30332:2

The advantage of natural gas as a feedstock is clearly demonstrated in these comparison because of the reduced investment cost and energy needs. It is only when gas prices become very high--say, \$5/MMBTU that coal becomes a competitor for ammonia production.

C. The Opportunity Cost of Natural Gas for Ammonia Production

8.05 There are many countries with reserves of natural gas that could be used for ammonia production. In some cases ammonia projects are not realized because the investment cost of recovering the gas and establishing a nitrogen complex is too expensive. In other cases one of the most important factors determining the feasibility of ammonia or urea production in any country will be the economic (opportunity) value of the particular gas resource available, or to be made available for such production. It is difficult to generalize about the economic value of gas because it varies from location to location depending on the size of the resource and on the alternative (opportunity) uses of the gas if it were not to be used for fertilizer production. If the gas can be used for oil substitution for example, then the economic value of the gas would be linked to the value of oil. However, in many countries, particularly developing countries, there are many occasions where this fuel oil substitution alternative is not available and where the economic value of gas is determined by other alternatives, in which gas has a lower value.

8.06 Alternative uses of gas include the fertilizer and petrochemical industry such as methanol, or LNG for overseas markets. Generally, because of their similar investment and processing costs the economics of methanol and ammonia are similar. Where deposits of gas are small, ammonia and urea plants are often the most attractive proposition. Where deposits are large, LNG manufactured may be feasible. Even so, the cost of liquifaction and transport of natural gas is expensive as can be seen from Fig. 22 which been adapted from [a] Opportunity costs for LNG manufacture will depend on the net-back of gas and hence on the corresponding oil price which determines this.

8.07 When oil is $30/BBL there is generally going to be a good case for LNG exports with net-backs equivalent to between $1.0-2.0/MMBTU. When oil is priced at $20/BBL transport becomes more important and there are few opportunities for LNG competing with oil and the construction of new facilities for LNG liquifaction and transport is unlikely to be viable any where. Because of the large investments needed together with uncertainty in future oil prices, export of LNG has not developed as was expected. Generally these factors have helped to reduce the opportunity costs for gas for ammonia manufacture which at most energy-rich locations is available at $1.0/MMBTU or less. In many cases, particularly where gas is being flared and has no apparent alternative immediate use, the opportunity cost for gas is basically that of collection and sweetening.

8.08 Although the opportunity cost of gas will set the lower level of gas price on which economic returns are calculated, in many cases the financial prices are often on a higher scale depending on what the market will accept.

---

[a] Akinobu Tsumura, "Perspectives on Natural Gas and LNG in the 1980s," Chemical Economy and Engineering Review, Japan, 1980, Volume 12, No. 6.

Figure 22

COST OF LIQUIFACTION AND TRANSPORTING NATURAL GAS

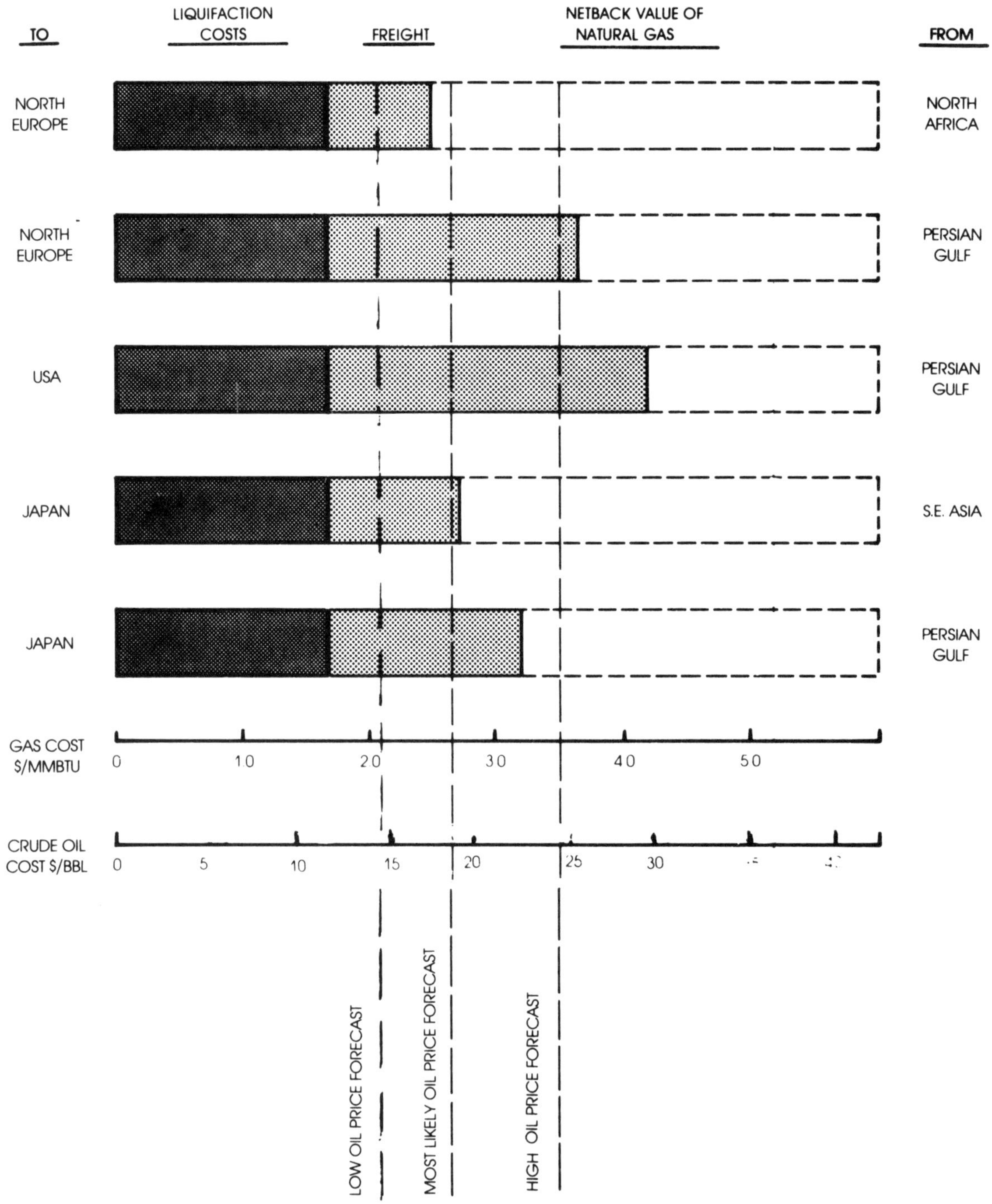

World Bank—30332:4

D. Discussion on Different Energy Pricing Scenarios

8.09 Undoubtedly one of the major factors constraining further investment in the fertilizer industry today is the uncertainty of future energy prices and hence fertilizer prices. This is particularly so in the USA and Western Europe where to an increasing extent, gas prices for ammonia production, are expected to become more and more linked to fuel oil prices. Certainly the outlook for these countries is much more favorable than it was several years ago when it was projected that both oil and gas prices would rise steadily to about $4-5/MMBTU by 1990. There was then considerable discussion on the possibilities of producing ammonia by coal gasification and with gas at these high prices the use of coal did not look unreasonable where it was available at less than $30/ton. Today the future outlook is somewhat different, with forecasts of oil and gas prices now falling within the range of $2.0-3.0/MMBTU and even less. With these forecasts, natural gas seems likely to remain the unrivalled feedstock for new plants commissioned during the next five years at least. Taking into account the impact of the reduced forecasts for gas prices and making allowances also for the most recent improvements in energy savings in new ammonia plants, the variable costs for ammonia production for the Developed Market Economies are now about $60-70/ton lower than forecasts made a few years ago. On the other hand, for a plant on a new site near a cheap gas source in a Developing Market Economy, the corresponding projected savings are much lower, perhaps $5-10/ton of ammonia. There is no longer such an obvious case for all new ammonia plants going where the gas is very cheap, particularly if it is an export-based plant and freight costs to the market are relatively high.

8.10 In the following discussion three different scenarios of energy pricing and their impacts on the world nitrogen industry are discussed. Oil prices are presented in 1986 constant US dollars. The high oil cost scenario assumed $24/BBL, the most likely cost scenario $18/BBL and the low cost scenario, $14/BBL.

1. Oil Prices of $14/BBL

8.11 A crude oil price of $14/BBL is equivalent to an energy price for gas of about $2.0/MMBTU. The obvious impact of gas at this price would be to make ammonia production much more attractive in those areas of the world, particularly the Developed Market Economies which have gas resources, competing with oil as a fuel and feedstock, such as the USA and Western Europe.

8.12 USA. Although this level of energy price is not much lower than average gas price to the US ammonia industry in 1986, there is a wide range of gas prices in the industry with about 20% of U.S. gas to ammonia plants above $3.5/MMBTU. The high cost producer in the USA is particularly vulnerable to cheap ammonia imports based on cheap natural gas in a period of oversupply and will continue to remain so. If US producers felt that gas prices could stay at a low level, there would be much more consideration of new investment. With relatively low capital charges and cash costs of perhaps $80/ton, a new ammonia plant in the USA could be in a relatively strong position to compete with imports from Canada or the Caribbean areas. However there is still considerable doubt in the US industry about a major sustained fall in energy prices and this is obviously going to constrain new investments even if oil prices were to remain at $14/BBL in the short term.

8.13 Western Europe. Generally the situation is very similar to that for the USA except that only a few countries have domestic gas under favored conditions but even so, the average level of gas prices at the present time, although it has fallen, is still probably above that in the USA. As in the USA, low energy price would very much strengthen the position of domestic production against exports and new capacity to replace worn-out plant would be more easily justified. The recent announcement of ICI, one of the world's largest ammonia producers to replace two old ammonia plants at its Sevenside Works in the UK with new capacity shows that at least one producer feels that with newer and improved technology and reasonable prospects for gas prices it can compete against imports in the future. Other producers have expressed similar views and some have been willing to replace existing plants during the last few years when the prospects for energy were not so favorable as they are today. Generally, however, it is not expected that the lowering of gas prices in Western Europe will do very much to change the current trend to reduce nitrogen exports and so long as USSR maintains a strong ammonia export potential to Western Europe, there will be little change in the position there.

8.14 Japan. This country has been affected more than any other developed country, in suffering from high priced feedstock to its ammonia plants. There has been a major retraction in the industry and Japan is now basically only providing ammonia to meet its own needs. Apart from fertilizers, a large part of ammonia in Japan goes to industrial use. Japan has recently converted most of its naphtha based plants to LNG and LPG and has now considerable flexibility with regard to feedstock source either as LPG, LNG and naphtha. A sustained fall in oil prices however is not likely to have a significant impact on the current Japanese ammonia industry other than to make naphtha a more attractive feedstock than LPG and to diminish the relative economic advantage of the large coal based plant recently commissioned.

8.15 Developing Market Economies with Cheap Gas. For most existing plants in these countries, a lowering of oil prices is not likely to have any major impact on the operating costs. Gas prices are already likely to be low and below an equivalent oil price of $14/BBL and not a major part of total production costs. On the other hand for countries with large reserves of gas who have been considering investing for the LNG market, a major fall in world oil prices would seriously diminish the net-back value from LNG prices almost certainly to a point where it is no longer attractive, and when ammonia and other chemicals would be more attractive projects. One major impact of lower energy prices is that more replacement capacity and even new capacity may continue to be built in the Developed Market Economies. This will mean less new capacity will be needed from other sources. In this situation the major hurdle for a Developing Country to enter the international nitrogen industry becomes increasingly the cost of establishing infrastructure and operating experience that will enable plants to operate at high rates and thus recover their full capital charge. Once this has been done, a second and third plant on an existing site look progressively more attractive. For the more wealthy energy-rich countries who have established industrial estates (for example the establishment of Al Jubail in Saudi Arabia) the prospects of building up a major export business in nitrogen fertilizers may be worthwhile. Other countries, who have cheap gas resources but have to provide the cost of large supporting infrastructure and where the plant will be completely export-based may be well advised to stay out of the business for the time being.

8.16 Centrally Planned Economies. Experience over the last few years shows that the behavior of these countries is often as much determined by political as well as economic consideration and this makes it difficult to predict what will happen there in the future. If energy prices continue to fall and the USA and Western Europe, production becomes more competitive against USSR imports it seems unlikely that there will be any new export-based investment in Eastern Europe. China is an importing country, slowly moving to self sufficiency and lower energy costs are not likely to affect it's situation very much other than to reduce the cost of its imports somewhat.

2. <u>Oil Prices of $18/BBL</u>

8.17 At the present time this is regarded as the most likely scenario for the period 1990-1995 and the analysis that has been made in Section X is based on this level of energy prices. Generally the situation will be similar to that discussed in scenario (a). With an equivalent gas price of $2.5/MMBTU, there will still be enough incentive for some developed countries to build replacement of new ammonia capacity. There will still not be sufficient incentive to encourage a significant increase in the LNG market. Nor is there sufficient incentive to go to other feedstocks like fuel oil, naphtha, or coal. When there is a situation of excess ammonia capacity, prices can fall to very low levels governed by the cash costs of production in the main importing countries. In a tighter situation, prices will be determined to some extent by the ability of the USA (the balancing market) to compete with new capacity against imported material based on low cost gas as indicated in Annexes 17 and 18.

3. <u>Oil Prices of $24/BBL</u>

8.18 This particular scenario relates to the energy situation a year or so ago and one to which a large extent still conditions thinking in the nitrogen industry today, particularly the USA. A sharp rise in energy prices over the next few years is likely to cause a slow-down in world economic growth which would probably lead to a fall in fertilizer demand. The reduced ability of many developing countries to purchase both food and fertilizer supplies would lead to some reduction in demand particularly as fertilizer prices would go up as a result of increased energy costs. This type of scenario would favor the building of new plants near the cheap gas sources and would probably prevent any further construction in the major Developed Market Economies.

8.19 <u>USA</u>. If oil prices increase to $24/BBL the equivalent cost for natural gas would be $3.5/MMBTU. Although gas prices would probably lag behind fuel oil prices until at least 1990 there would be a gradual closing of the gap as new gas contracts are made at higher price levels. By 1990 new gas prices would be about $3.5/MMBTU and at that price few producers would consider new capacity in the USA. There could even be a further retraction in the nitrogen industry there as worn-out plants are not replaced. Many producers would look to further imports of ammonia and urea from Canada and the Caribbean. Also, the increased cost of energy, increased interest rates, could lower the growth of US agricultural production resulting in a slow-down in nitrogen fertilizer demand and US imports of ammonia. This would reduce the prospects for new plants which are looking to the USA as their major market.

8.20 Western Europe. The situation here would be similar to that in the USA. There would be few prospects of new ammonia plants being built or replaced and the region would become increasingly dependent on imports. The exports of finished fertilizers would decline more rapidly.

8.21 Developing Market Economies. Those countries with resources of oil and gas would obviously benefit from higher oil prices. The increased revenues in some major countries like Mexico and Nigeria would stimulate fertilizer demand and also provide investment opportunties for new ammonia and urea plants. Nitrogen plants in these countries would become increasingly competitive with plants in the Developed Market Economies particularly once an infrastructure has been established.

8.22 Centrally Planned Economies. Most of the energy resources of Eastern Europe are in the USSR. If the price of oil were to increase significantly the opportunity price of Russian gas would also increase both as an export material to Europe and also as a feedstock for nitrogen fertilizers. The USSR has however had a bad experience of investing in fertilizer plants and may be reluctant to repeat this experience. A major increase in energy prices is not likely to have much impact on the Chinese nitrogen situation. China imports about 20% of its nitrogen needs but its nitrogen consumption is approaching maturity and it will become more self sufficient as it develops its own resources.

8.23 Overall View. For the last decade there has been a strong move towards nitrogen production in countries where the gas is very cheap mainly in the Developing Market Economies and Centrally Planned Economies. Almost all the export nitrogen trade is based on cheap gas and the experience of the last few years has indicated that only those companies, who have very low cash costs, can survive. The experience in the first half of the 1980s which closed down about 6 million tons of capacity in the Developed Market Economies will continue to remain a major constraint to future building in this region even if there is a decrease in energy prices. Also there is not at this stage sufficient confidence in a sustained period of low energy prices that would justify major new investments. Although changes in oil prices in the future will obviously have an impact on energy prices to nitrogen plants in the Developed Market economies, it will have little impact on the opportunity cost of gas to those plants which already dominate the export market. It is expected therefore that the present trend of building export-based plants where the gas is cheap will continue, although if oil prices decline, the trend will slow down, or accelerate if oil prices increase. The uncertainty of future energy prices together with other constraining factors such as the world depressed agricultural situation will have a delaying effect on new investments which could exacerabate a short fall in supply in the early 1990s.

IX. WORLD NITROGEN FERTILIZER CONSUMPTION AND FOOD PRODUCTION

A. General

9.01 Before examining the future prospects of the world nitrogen industry in terms of food production, it is useful to review the major historical changes that have taken place so far in the industry. Between the turn of the century and about 1950, world nitrogen consumption increased on average at the relatively modest rate of about 5% py. Between 1950 and 1970, nitrogen fertilizer consumption increased at an average yearly rate of about 10% and between 1970 and 1980 at a rate of 7%. Between 1980 and 1985 however the growth rate declined to 4% and for the next decade is projected to grow at less than 3% py.

9.02 These projected growth rates are significantly less than analysts were making several years ago. What is not too clear at this stage is whether or not current projections are being too influenced by the current world recession particularly in the agricultural industry in the early 1980s. However, even though fertilizer consumption has fallen off significantly in the last few years, there has been as yet, no apparent corresponding fall-off in the growth rate of food production. FAO estimates of the 1985/86 cereal crop is 1,835 million tons and forecasts for the 1986/87 crop are 1,828 million tons only slightly below the record figure of 1985/86. In the Developed Market Economies of North America and Western Europe, a surplus of agricultural products is forcing governments to restrict food production and this is constraining fertilizer consumption. Many major developing countries which several years ago were forecast to remain major importers of food for many years to come, have in the last year or two become grain exporters. World cereal stocks are forecast to reach a record 433 million tons by the end of the 1986/87 season which is 39 million tons more than the previous year. Forecasts of world cereal trade have been reduced further to reflect the favorable harvest outlook in the major importing countries.

9.03 This situation does not mean that all people in these countries are receiving all the food they need, due to constraints in income and food distribution, but nevertheless there does appear to be a more favorable food situation in many developing countries that was foreseen a few years ago. Undoubtedly, favorable weather and crop growing conditions in the last few years has played a part in increasing food production. So also, has the continuing development and use of higher yielding varieties which together with improved soil management irrigation and improved water management, the use of agrochemicals and probably a more efficient use of fertilizers at existing rates are all playing a part in increasing food production.

B. Food and Nitrogen Fertilizer Forecasts

9.04 In order to establish a relationship between nitrogen fertilizer consumption and food production, use has been made of the time series for cereal grain production and time series for nitrogen fertilizer production from the FAO production Year Books. Although the FAO food production index could also be used as an indication of food production, grain production is preferred as a single tangible product rather than an index involving many products. Cereal grains as a group are by far the most important component of the world's food supply receiving more than half of all fertilizers used. Because of the major position of grains in food production there has always been a high degree of correlation between the FAO food production index and grain production for the world as a whole and for most regions. The time series information for population, grain production and nitrogen consumption is given in Annex 19 and the following relationships utilized in the regression models.

1. Grain Production (G) against Time (T)

9.05 The best model obtained was of the form Log G = A+BT with an R^2 a/ of 0.99. The projections developed by this model are found to be in close agreement with the relationship developed by UNIDO in similar studies in the late 1970s. Plots of actual grain production (G) against Time (T) and projected values based on Log G against T are shown in Fig. 23. Grain production projections through the year 2000 are given in Table 25. Grain production per capita projections have been derived assuming a population growth rate of 1.55% and these are also given in table 25. This indicates that if the present trend continues there will be a slight improvement in grain production per capita, equivalent to about 1% py through the year 2000.

2. Fertilizer Consumption (N) against Time (T)

9.06 Several relationships were examined and the most adequate models were of the form.

(a) $N = A+BT+CT^2$ ($R^2 = 0.99$)

(b) $N = A+BT$ ($R^2 = 0.98$)

A plot of actual and forecasted values using model (a) is shown in Fig. 24. The projections of nitrogen consumption through the year 2000 for (a) and (b) are shown in table 26.

3. Grain Production per Unit of Nitrogen $\left(\frac{G}{N}\right)$ against Time (T)

9.07 Various functional forms were explored to forecast $\left(\frac{G}{N}\right)$ against (T). These included log transformations, reciprocal transformations, and weighted least squares procedures. A weighted least squares procedure for the model $\left(\frac{G}{N}\right)$= A+BT was utilized. This procedure gave greater weightage to the more

---

a/ R - square is the portion of variability explained by the model, e.g., R - square of 1.0 indicates a perfect fit with no errors.

recent observations and heavily discounted the distant past values. The projected values indicated a relative flattening of the curve in the early 1980s and guided by these results a logistic model of the following functional form was also tried.

$$\left(\frac{G}{N}\right) = A[1 + \exp.(B+CT)]$$

A log transformation converted this into a linear relationship of the form

(c) $$\text{Log}\left[\frac{G}{NA} - 1\right] = B+CT$$

A high R^2 was obtained from this regression model and a plot of actual and forecasted values is given in Fig. 25 and in Table 25. Nitrogen requirement to maintain the projected grain production was derived from the projected $\left(\frac{G}{N}\right)$ ratio and are given in Table 26.

4. Discussion of Results

9.07 Although future grain production will undoubtedly depend on increased nitrogen fertilizer use, it is appreciated that there are many other factors which are also increasing production such as increased land under cultivation, high yielding varieties, irrigation and other fertilizer nutrient use etc. However the fact that the $\left(\frac{G}{N}\right)$ ratio is now levelling indicates that the impact of many of these factors are no longer dynamic. It is possible, however, that if some major breakthroughs were to be made in other inputs such as biological fixation or growth regulators, the rate of $\left(\frac{G}{N}\right)$ could in fact go up again in time. Nevertheless this time series analysis is of interest for generally it indicates that if grain production continues the upward trend as indicated in series (c) nitrogen production will have to increase to about 139 million tons by the year 2000 - a growth rate of about 4.3% py. On the other hand if the long-term trend of nitrogen against time is to continue, according to series (b)

Figure 23

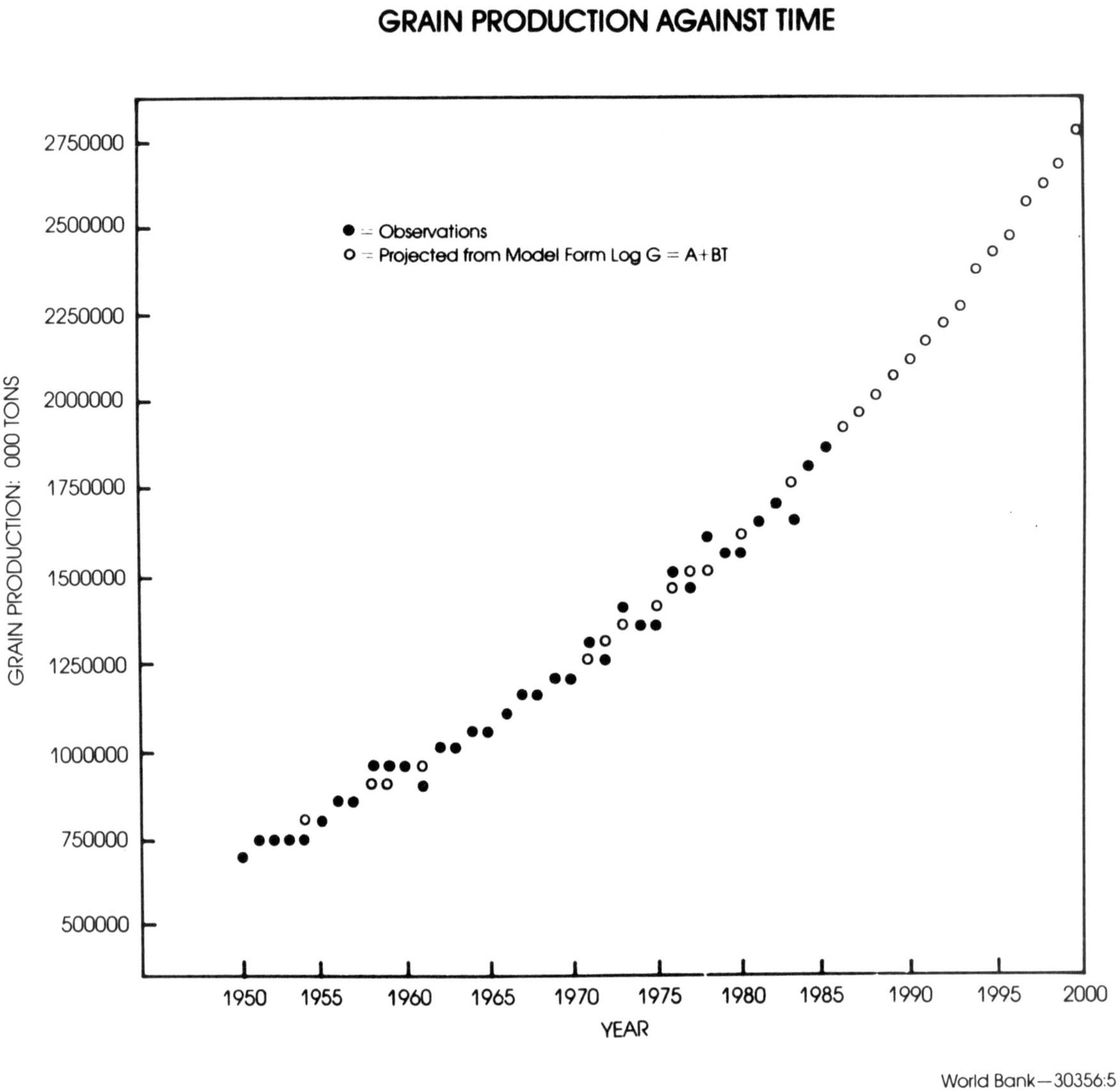
GRAIN PRODUCTION AGAINST TIME
● = Observations
○ = Projected from Model Form Log G = A+BT
GRAIN PRODUCTION: 000 TONS
2750000
2500000
2250000
2000000
1750000
1500000
1250000
1000000
750000
500000
1950
1955
1960
1965
1970
1975
1980
1985
1990
1995
2000
YEAR
World Bank—30356:5

Figure 24

NITROGEN CONSUMPTION AGAINST TIME

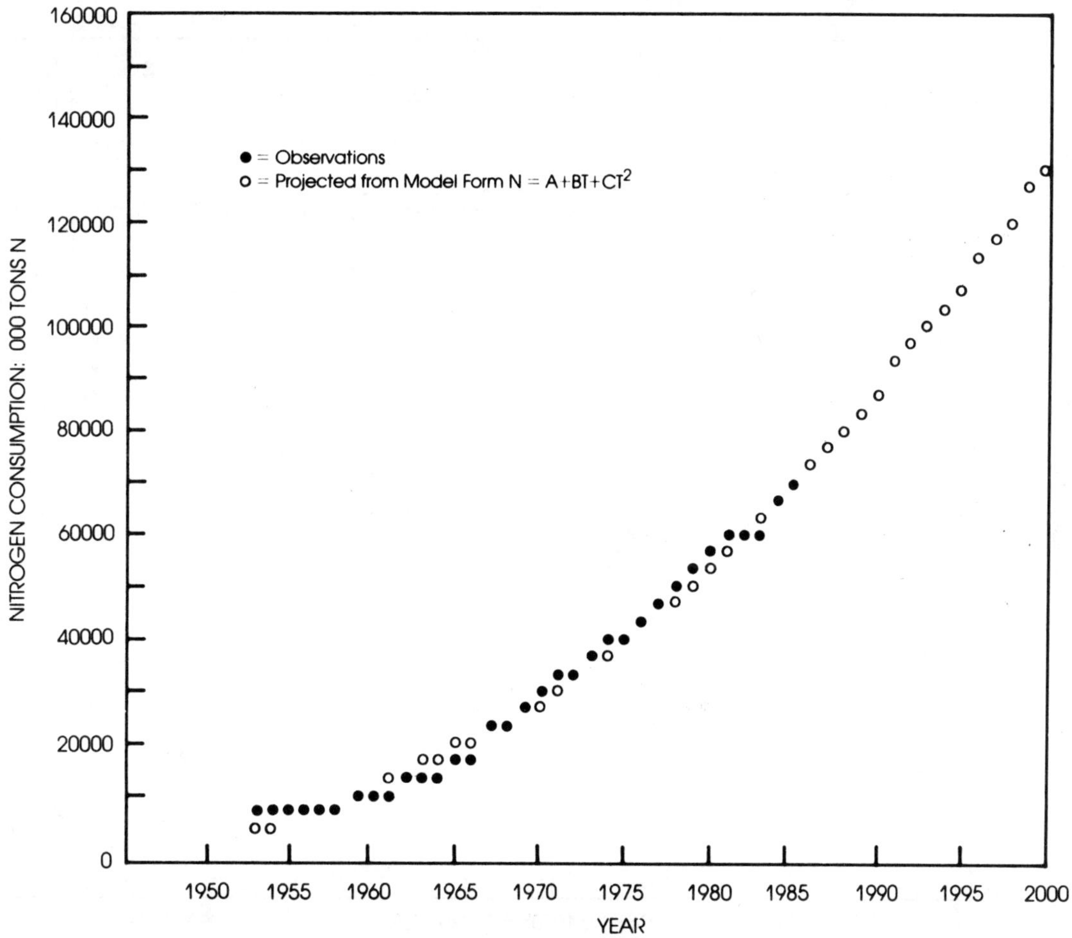

World Bank—30356:6

Figure 25

GRAIN PRODUCTION NITROGEN CONSUMED RATIO AGAINST TIME

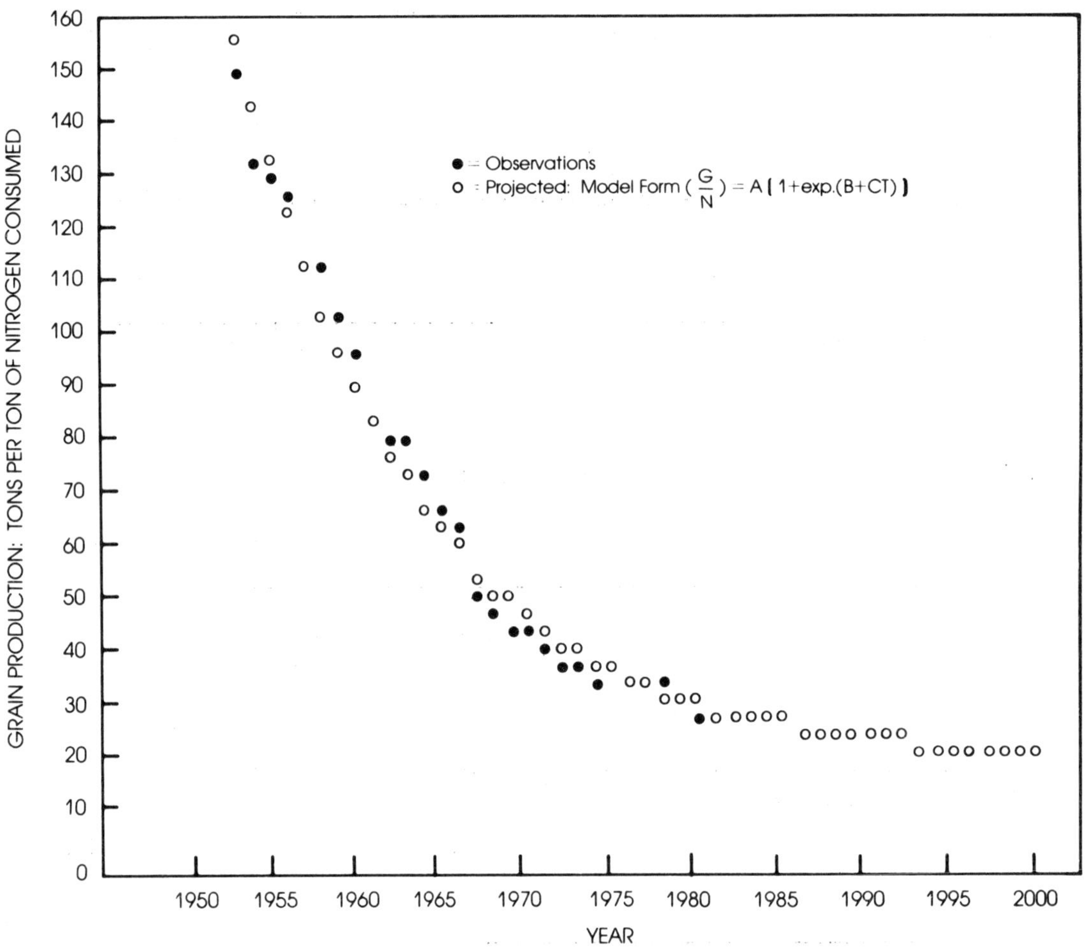

World Bank—30356:7

nitrogen consumption will reach 130 million tons by the year 2000 equivalent to a growth rate of 3.9% py. Both of the figures are higher than the effective demand of about 2.5-3% being projected by most analysts at the present time and it does raise the question of whether current forecasts are being too influenced by the short-term depression in the agricultural industry. On the other hand, the projection of the existing time series does indicate an increase in food production per capita of about 20% through 2000. If no increase in grain production per capita is assumed and if it is further assumed that grain production per ton of nitrogen is basically a function of total grain production then projected nitrogen demand by 2000 would be 108 million tons N representing an average annual demand increase of 2.7%.

9.08 Generally this very brief exploratory look at the relationship between food production and nitrogen fertilizer, summarized above, can only be indicative but they do concur with the results of the FAO study "Agriculture to the Year 2000", that if a modest improvement in the economic welfare of the developing countries is to be achieved these countries will have to increase their fertilizer use by about 7% py through the year 2000. This compares with the Working Group's projections of a growth rate in effective demand of about 5% py in the Developing Market Economies. Further work is obviously needed to establish more precisely the relationship between future food production and fertilizer needs.

Table 25

Time Series Projections of World Grain Production

Year	Projected World Population (Million)	Projected Grain Production (Million Tons)	Grain production per Capita (Tons)	Grain Production per Ton Nitrogen (Tons)
1984	4,784	1,801	0.376	25.5
1990	5,246	2,095	0.399	22.3
1995	5,666	2,397	0.423	20.7
2000	6,119	2,742	0.445	19.7

Table 26

Time Series Projections of Nitrogen Fertilizer Demand
(Million tons N)

Year	From Nitrogen Time Series (b)	From Nitrogen Time Series (a)	Derived From Cereal Time Series	Working Group Projections	No Increase in Food Production per Capita
1984	70.5	70.5	70.5	70.5	70.5
1985	69.6	69.6	69.6	69.6	
1990	88.0	90.9	93.9	80.5	
1995	108.1	112.7	115.7	92.2	
2000	130.3	136.8	139.2	NA	108.0
Demand Increase % py	3.9	4.2	4.3	2.6	2.7

5. The World Bank Integrated Agricultural/Fertilizer Model

9.08 In order to examine the longer term demand for fertilizers in a more systematic way the World Bank Economic and Projections Department has developed an integrated agricultural/fertilizer model that derives world fertilizer demand explicitly from that of grain production. In the model, the demand for fertilizers is determined by, among other things, the market clearing prices and production and consumption of grains. The model also includes factors such as other agricultural inputs, growing population income and cyclical investments in new capacity.

9.09 A critical factor in the short-term outlook for nitrogen fertilizer demand is the likely impact of the U.S. Farm Bill which came into effect in December 1985. The bill in essence sharply reduced the farm price support in exchange for higher income support and provides measures for acreage reduction. The bill is also likely to have an effect on both global crop prices and fertilizer demand. Simulations with the integrated agricultural/fertilizer model indicate that the effect of the acreage reduction program will result in a

9.7% reduction in U.S. fertilizer demand (11.8% decline for nitrogen, 5.8% for phosphate and 9.4% for potash) while it will have a small positive impact on the rest of the world's fertilizer consumption.

9.10 Over the medium to longer term, the model projects total world fertilizer demand to grow at 2.6% per year between 1984/85 and 1989/90, and at 2.6% per year between 1989/90 and 1999/2000. As in the past, nitrogenous fertilizer consumption is projected to increase at a faster rate than consumption of phosphate and potash but the difference in growth rates are expected to narrow considerably as nitrogen application begins to fall off in those countries where it is already heavily used such as China. The regional breakdown of nitrogen demand growth rate determined by the World Bank integrated agricultural/fertilizer model is compared with the projections of the Fertilizer Working Group in Table 27. There are some differences in the time period and regional classifications but they are regarded as sufficiently similar to make valid comparisons.

Table 27

Long Term Nitrogen Fertilizer Demand
(Average % Yearly Demand Increase)

	EPD Model 1985-2000	Working Group 1985-1996
North America	1.3	0.6
EEC	1.1	
Western Europe		1.1
E. Europe	2.0	3.3
Asia	4.1	
Near East		3.6
Far East		5.5
World .	2.9	2.6

The results from the World Bank EPD model are generally in reasonable agreement with the Working Group's projections.

X. FUTURE OUTLOOK FOR THE WORLD NITROGEN INDUSTRY

A. Potential Nitrogen Projects to Meet Future World Nitrogen Needs

10.01 An attempt has been made to forecast regions and countries in which investment may occur, to bring nitrogen fertilizers plants onstream after 1990/91, to help meet the world's nitrogen needs through 1995/96. As indicated earlier, about 56 new or refurbished plants will be required throughout this period, and possibly more, as more plants need replacing. In this exercise it is assumed that implementation of new projects to meet export and domestic needs will be determined mainly on economic grounds based on realization prices from the data contained in Annexes 17 and 18. Factors other than realization price have however been taken into account, for example the technical and financial resources that would be needed to implement one or more projects, together with a country's ability to supply these. Market location and freight costs will often play an important part in determining whether or not a project can compete successfully. The list of projects given in table 28 is a value judgement assessment and should be used as a guide only. The following assumptions have been made in making this assessment:

(a) The Developed Market Economies and many of the Developing Market Economies after 1990 will have a growing deficit of nitrogen fertilizers and will depend mainly on exports from the USSR and other Eastern European countries. The USSR has many technical problems with its plants, has a growing domestic need for nitrogen fertilizers and does not appear to have any major new investment programs for nitrogen similar to that which took place in the 1970's. Only a few new plants are expected to be built in the USSR and therefore most new plants to meet increasing demand will have to be built in the Market Economies.

(b) Three major markets for nitrogen exports in the future will be the Far East (including China), Western Europe and the USA. Ammonia will be the main product demanded by USA and Western Europe and those developing countries with phosphate processing facilities such as Morocco, Tunisia, Jordan, etc. Urea will be the main export material in the Far East.

(c) The most attractive locations for plants to meet the needs of the USA, and Latin American market will be Trinidad, Venezuela, Mexico, and Argentina. Western European markets will continue to be served by USSR and other Eastern European countries but new plants in Algeria and Libya should also be able to compete in Western Europe.

(d) The export needs of the Far East Market will continue to diminish as many countries with gas such as China, Indonesia, Thailand, Pakistan, Bangladesh, etc. continue to build their own plants to meet their own needs and in some cases provide material for export. The Near East will continue to increase its export capacity to meet its own needs and demand from S.E. Asia and the Far East.

(e) Although there are several potential locations for new plants in Africa where gas is available cheaply (Nigeria, Tanzania, Mozambique, Cameroon, etc.) because investment costs and freight costs are likely to be high, only those plants with large domestic consumption or relatively close captive markets are likely to be attractive. The plants referred to in Table 27 are new plants and additional to those already announced to come on stream before 1990/91. The average size is assumed to be equivalent to 1,350 metric tpd ammonia plant capacity operating at 90% utilization.

B. Ammonia and Urea Requirements to Meet Future Nitrogen Demand

10.02 The projects in Table 28 refer to nitrogen nutrients only and do not distinguish between ammonia and urea plants. Although it is normal to assess world nitrogen needs in terms of the nitrogen nutrient N it is clear that the nitrogen market is made up of two components, the ammonia market and the urea market. Almost all nitrogen fertilizers are produced from ammonia so there are basically no problems in assessing the needs of nitrogen in terms of ammonia capacity. Whether this ammonia is converted into urea should depend specifically on the demand for urea. From time to time the ratio of ammonia to urea capacity appears to be out of balance to meet world needs. When the ratio of urea: ammonia prices exceeds unity there is probably an imbalance in favor of ammonia capacity whereas when the ratio falls well below unity there is an excess of urea compared with ammonia capacity.

10.03 The supply demand balances derived in Section III imply that there will be no major change in international trading patterns for nitrogen fertilizers and intermediates in the next five years. The Developed Market Economies will continue to be the major importers of ammonia and the Developing Market economies together with Centrally Planned Asia will be the main importer of urea. Eastern Europe will remain the main exporter of both ammonia and urea. In 1984, roughly equal quantities of nitrogen in the form of urea and ammonia were traded. New producers entering the market will have to look carefully at which of these markets they are best able to compete. Generally, new plants in Europe, North Africa, North America and the Caribbean will tend to compete in the ammonia markets whereas plants in the Near and Far Easts in the urea markets.

Table 28

POTENTIAL NITROGEN PROJECTS TO MEET FUTURE NITROGEN NEEDS 1990/91-1995/96

Region	Country	No. of Plants	Capacity (Million tons N)
North America		4	1.48
	Canada	2	
	USA	2	
West Europe		2	0.74
Oceania		1	0.37
	Australia	1	
Africa		6	2.22
	Algeria	1	
	Egypt	1	
	Libya	1	
	Nigeria	1	
	Others	2	
Latin America		10	3.70
	Argentina	1	
	Brazil	1	
	Chile	1	
	Mexico	2	
	Trinidad	2	
	Venezuela	1	
	Others	2	
Near East		7	2.59
	Iran	1	
	Kuwait	1	
	Qatar	1	
	Saudi Arabia	2	
	Others	2	
Far East		12	4.44
	Bangladesh	1	
	Burma	1	
	Indonesia	2	
	India	5	
	Others	3	
East Europe		8	2.96
	USSR	6	
	Others	2	
Socialist Asia		6	2.22
	China	5	
	Others	1	
TOTAL		56	20.72

C. Conclusions

10.04 The world fertilizer industry is currently suffering what must be its worst recession ever, reflecting a depressed world agricultural industry which has lowered demand and prices of nitrogen fertilizers. This situation will probably continue for another two years or so but even with a relatively modest average world increase in demand for Nitrogen of 2.6% py, a balanced situation will be reached by 1990/91. The rate of building of ammonia capacity has fallen off significantly in recent years in a situation of excess supply, and forecasts for additional new construction are also much reduced. To avoid a serious shortage in the early 1990s at least eleven large new or refurbished plants will need to come on stream each year after 1990/91. Some of these plants will be needed to replace some of the worn-out plants built during the 1960s.

10.05 The changing prospects for energy prices and nitrogen trading patterns will mean that greater care than ever will have to be taken to ensure that only those projects offering reasonable and worthwhile return to their owners are implemented. The World Bank and other international lending agencies will have an increasing responsibility to the developing countries to see that this objective is achieved.

ANNEX 1. Regional and World Ammonia Plant Capacity ('000 Tons N)

	1984/85	1985/86	1986/87	1987/88	1988/89	1989/90	1990/91
DEVELOPED M.E.	35,346	35,265	35,609	35,784	36,015	36,380	36,745
North America	16,504	16,794	17,183	17,508	17,508	17,508	17,508
Western Europe	15,299	15,466	15,421	15,271	15,502	15,867	15,867
Oceania	525	525	525	525	525	525	890
Other Developed M.E.	3,018	2,480	2,480	2,480	2,480	2,480	2,480
DEVELOPING M.E.	21,613	23,888	24,486	25,443	27,359	29,322	31,048
Africa	404	747	1,019	1,019	1,291	1,291	1,291
Latin America	5,833	5,833	5,833	5,833	6,203	6,620	7,235
Near East	4,286	4,884	4,884	5,374	5,646	5,918	6,299
Far East	11,090	12,424	12,750	13,217	14,219	15,493	16,223
CENTRALLY PLANNED EC.	51,075	52,571	53,406	54,654	54,654	54,654	54,654
Eastern Europe	33,467	34,419	34,789	34,993	34,993	34,993	34,993
Socialist Asia	17,608	18,152	18,617	19,661	19,661	19,661	19,661
WORLD	108,034	111,724	113,501	115,881	118,028	120,356	122,447

Regional and World Ammonia Plant Capacity ('000) (Continued)

	1984/85	1985/86	1986/87	1987/88	1988/89	1989/90	1990/91
DEVELOPED M.E.	35,346	35,265	35,609	35,784	36,015	36,380	36,745
North America	16,504	16,794	17,183	17,508	17,508	17,508	17,508
Canada	2,909	3,215	3,508	3,806	3,806	3,806	3,806
U.S.A.	13,595	13,579	13,675	13,702	13,702	13,702	13,702
Western Europe	15,299	15,466	15,421	15,271	15,502	15,867	15,867
Austria	411	411	411	411	411	411	411
Belgium-Lux.	573	474	474	474	474	474	474
Finland	67	67	67	67	67	67	67
France	2,467	2,467	2,467	2,467	2,467	2,832	2,832
Germany, F.R.	2,032	2,032	2,032	2,032	2,032	2,032	2,032
Greece	266	376	331	331	331	331	331
Iceland	12	12	12	12	12	12	12
Ireland	365	365	365	365	365	365	365
Italy	1,722	1,722	1,722	1,722	1,722	1,722	1,722
Netherlands	2,719	2,719	2,719	2,719	2,786	2,786	2,786
Norway	647	647	647	647	647	647	647
Portugal	394	394	394	244	244	244	244
Spain	811	720	720	720	720	720	720
Sweden	57	57	57	57	57	57	57
Switzerland	53	53	53	53	53	53	53
United Kingdom	1,758	1,758	1,758	1,758	1,922	1,922	1,922
Yugoslavia	945	1,192	1,192	1,192	1,192	1,192	1,192
Oceania	525	525	525	525	525	525	890
Australia	449	449	449	449	449	449	814
New Zealand	76	76	76	76	76	76	76
Other Developed M.E.	3,018	2,480	2,480	2,480	2,480	2,480	2,480
Israel	68	68	68	68	68	68	68
Japan	2,260	1,722	1,722	1,722	1,722	1,722	1,722
South Africa	690	690	690	690	690	690	690

(Table continued on the following page)

Regional and World Ammonia Plant Capacity ('000 Tons N) (Continued)

	1984/85	1985/86	1986/87	1987/88	1988/89	1989/90	1990/91
DEVELOPING M.E.	21,613	23,888	24,486	25,443	27,359	29,322	31,048
Africa	404	747	1,019	1,019	1,291	1,291	1,291
Algeria	272	544	816	816	816	816	816
Malagasy Republic		46	46	46	46	46	46
Nigeria					272	272	272
Somalia		25	25	25	25	25	25
Zambia	82	82	82	82	82	82	82
Zimbabwe	50	50	50	50	50	50	50
Latin America	5,833	5,833	5,833	5,833	6,203	6,620	7,235
Argentina	78	78	78	78	78	126	126
Brazil	1,025	1,025	1,025	1,025	1,025	1,025	1,271
Chile							
Colombia	150	150	150	150	150	150	150
Cuba	321	321	321	321	321	321	321
Mexico	2,468	2,468	2,468	2,468	2,468	2,837	3,206
Peru	116	116	116	116	116	116	116
Trinidad	1,024	1,024	1,024	1,024	1,394	1,394	1,394
Venezuela	651	651	651	651	651	651	651
Near East	4,286	4,884	4,884	5,374	5,646	5,918	6,299
Abu Dhabi	272	272	272	272	272	272	272
Afghanistan	58	58	58	58	58	58	58
Bahrain		272	272	272	272	272	272
Egypt	966	966	966	966	966	966	1,347
Iran	38	364	364	364	364	364	364
Iraq	41	41	41	41	41	313	313
Kuwait	706	706	706	706	706	706	706
Libyan Arab Rep.	544	544	544	544	544	544	544
Qatar	488	488	488	488	488	488	488
Saudi Arabia	435	435	435	835	835	835	835
Sudan				90	90	90	90
Syria Arab Rep.	313	313	313	313	313	313	313
Turkey	425	425	425	425	697	697	697
Far East	11,090	12,424	12,750	13,217	14,219	15,493	16,223
Bangladesh	554	554	554	826	826	826	826
Burma	115	213	213	213	213	213	213
India	5,903	6,998	7,324	7,568	8,298	9,028	9,758
Indonesia	2,460	2,460	2,460	2,411	2,683	2,955	2,955
Korea Rep. of	734	734	734	734	734	734	734
Malaysia	41	313	313	313	313	313	313
Pakistan	1,136	1,152	1,152	1,152	1,152	1,152	1,152
Sri Lanka	147						
Thailand						272	272

(Table continued on the following page)

Regional and World Ammonia Plant Capacity ('000 Tons N) (Continued)

	1984/85	1985/86	1986/87	1987/88	1988/89	1989/90	1990/91
CENTRALLY PLANNED EC.	51,075	52,571	53,406	54,654	54,654	54,654	54,654
Eastern Europe	33,467	34,419	34,789	34,993	34,993	34,993	34,993
Albania	82	82	82	82	82	82	82
Bulgaria	966	1,238	1,238	1,238	1,238	1,238	1,238
Czechoslovakia	887	787	787	787	787	787	787
Germany, Dem. Rep.	1,318	1,318	1,318	1,318	1,318	1,318	1,318
Hungary	853	853	853	853	853	853	853
Poland	2,124	2,124	2,124	2,328	2,328	2,328	2,328
Romania	3,505	3,505	3,505	3,505	3,505	3,505	3,505
U.S.S.R.	23,732	24,512	24,882	24,882	24,882	24,882	24,882
Socialist Asia	17,608	18,152	18,617	19,661	19,661	19,661	19,661
China	16,394	16,938	17,238	18,282	18,282	18,282	18,282
Korea, D.P.R.	871	871	871	871	871	871	871
Taiwan	289	289	289	289	289	289	289
Vietnam, Dem. Rep.	54	54	219	219	219	219	219
WORLD TOTAL	108,034	111,724	113,501	115,881	118,028	120,356	122,447

Source: World Bank/FAO/UNIDO
Fertilizer Working Group
December 1986

ANNEX 2. Regional and World Ammonia Plant Supply Capability ('000 Tons N)

	1984/85	1985/86	1986/87	1987/88	1988/89	1989/90	1990/91
DEVELOPED M.E.	31,829	31,409	31,798	32,552	32,757	33,009	33,218
North America	15,158	14,994	15,228	15,983	16,165	16,208	16,222
Western Europe	13,465	13,506	13,874	13,861	13,884	14,093	14,288
Oceania	473	476	476	476	476	476	476
Other Developed M.E.	2,733	2,433	2,220	2,232	2,232	2,232	2,232
DEVELOPING M.E.	16,240	17,382	18,615	19,879	20,986	22,389	23,800
Africa	233	307	307	523	632	737	759
Latin America	4,785	4,838	4,850	4,850	4,998	5,300	5,699
Near East	3,135	3,399	3,931	4,160	4,457	4,646	4,864
Far East	8,087	8,838	9,527	10,346	10,899	11,706	12,478
CENTRALLY PLANNED EC.	41,140	42,219	43,189	44,083	44,687	44,823	44,876
Eastern Europe	26,135	27,081	27,701	28,034	28,181	28,211	28,219
Socialist Asia	15,005	15,138	15,488	16,049	16,506	16,612	16,657
WORLD	89,209	91,010	93,602	96,514	98,430	100,221	101,894

Regional and World Ammonia Plant Supply Capability ('000 Tons N) (Continued)

	1984/85	1985/86	1986/87	1987/88	1988/89	1989/90	1990/91
DEVELOPED M.E.	31,829	31,409	31,798	32,552	32,757	33,009	33,218
North America	15,158	14,994	15,228	15,983	16,165	16,208	16,222
Canada	2,687	2,859	3,122	3,390	3,559	3,602	3,616
U.S.A.	12,471	12,135	12,106	12,593	12,606	12,606	12,606
Western Europe	13,465	13,506	13,874	13,861	13,884	14,093	14,288
Austria	411	411	411	411	411	411	411
Belgium-Lux.	437	391	436	436	436	436	436
Finland	67	67	67	67	67	67	67
France	2,369	2,270	2,270	2,270	2,270	2,404	2,555
Germany F.R.	1,929	1,930	1,930	1,930	1,930	1,930	1,930
Greece	239	279	303	293	298	298	298
Iceland	11	11	11	11	11	11	11
Ireland	347	347	347	347	347	347	347
Italy	1,314	1,464	1,464	1,464	1,464	1,464	1,464
Netherlands	2,434	2,515	2,567	2,583	2,580	2,595	2,629
Norway	615	615	615	615	615	615	615
Portugal	186	85	286	251	200	189	183
Spain	724	651	612	612	612	612	612
Sweden	51	51	51	51	51	51	51
Switzerland	48	48	48	48	48	48	48
United Kingdom	1,705	1,705	1,705	1,705	1,769	1,840	1,856
Yugoslavia	578	666	751	767	775	775	775
Oceania	473	476	476	476	476	476	476
Australia	404	404	404	404	404	404	404
New Zealand	69	72	72	72	72	72	72
Other Developed M.E.	2,733	2,433	2,220	2,232	2,232	2,232	2,232
Israel	61	61	61	61	61	61	61
Japan	2,062	1,755	1,538	1,550	1,550	1,550	1,550
South Africa	610	617	621	621	621	621	621

(Table continued on the following page)

Regional and World Ammonia Plant Supply Capability ('000 Tons N) (Continued)

	1984/85	1985/86	1986/87	1987/88	1988/89	1989/90	1990/91
DEVELOPING M.E.	16,240	17,382	18,615	19,879	20,986	22,389	23,800
Africa	233	307	307	523	632	737	759
Algeria	136	190	170	381	401	408	408
Malagasy Republic		13	27	31	32	32	32
Nigeria					87	185	207
Somalia		5	11	12	13	13	13
Zambia	47	49	49	49	49	49	49
Zimbabwe	50	50	50	50	50	50	50
Latin America	4,785	4,838	4,850	4,850	4,998	5,300	5,699
Argentina	66	66	66	66	66	83	101
Brazil	923	962	974	974	974	974	1,067
Chile							
Colombia	105	105	105	105	105	105	105
Cuba	161	161	161	161	161	161	161
Mexico	1,974	1,974	1,974	1,974	1,974	2,092	2,343
Peru	58	58	58	58	58	58	58
Trinidad	1,010	1,024	1,024	1,024	1,172	1,339	1,376
Venezuela	488	488	488	488	488	488	488
Near East	3,135	3,399	3,931	4,160	4,457	4,646	4,864
Abu Dhabi	227	253	267	267	267	267	267
Afghanistan	44	44	44	44	44	44	44
Bahrain		103	220	245	258	258	258
Egypt	725	725	725	725	725	725	839
Iran	19	84	158	174	182	182	182
Iraq	19	21	21	21	21	75	136
Kuwait	299	333	658	671	671	671	671
Libyan Arab Rep.	451	462	462	462	462	462	462
Qatar	488	488	488	488	488	488	488
Saudi Arabia	400	413	413	565	736	774	793
Sudan				22	46	51	54
Syria Arab Rep.	157	157	157	157	157	157	157
Turkey	306	316	318	319	400	492	513
Far East	8,087	8,838	9,527	10,346	10,899	11,706	12,478
Bangladesh	393	410	414	497	589	609	620
Burma	64	110	147	156	160	160	160
India	4,237	4,532	5,026	5,488	5,855	6,351	6,881
Indonesia	1,756	2,092	2,183	2,196	2,274	2,478	2,611
Korea Rep. of	439	439	439	661	661	661	661
Malaysia	35	127	231	254	266	266	266
Pakistan	1,060	1,077	1,087	1,094	1,094	1,094	1,094
Sri Lanka	103	51					
Thailand						87	185

(Table continued on the following page)

Regional and World Ammonia Plant Supply Capability ('000 Tons N) (Continued)

	1984/85	1985/86	1986/87	1987/88	1988/89	1989/90	1990/91
CENTRALLY PLANNED EC.	41,140	42,219	43,189	44,083	44,687	44,823	44,876
Eastern Europe	26,135	27,081	27,701	28,034	28,181	28,211	28,219
Albania	74	74	74	74	74	74	74
Bulgaria	913	1,021	1,137	1,163	1,176	1,176	1,176
Czechoslovakia	816	770	724	724	724	724	724
Germany, Dem. Rep.	1,186	1,186	1,186	1,186	1,186	1,186	1,186
Hungary	810	810	810	810	810	810	810
Poland	1,601	1,675	1,691	1,764	1,838	1,854	1,862
Romania	2,937	2,969	2,979	2,979	2,979	2,979	2,979
U.S.S.R.	17,798	18,576	19,100	19,334	19,394	19,408	19,408
Socialist Asia	15,005	15,138	15,488	16,049	16,506	16,612	16,657
China	13,935	14,120	14,430	14,946	15,394	15,495	15,540
Korea, D.P.R.	784	784	784	784	784	784	784
Taiwan	254	202	202	202	202	202	202
Vietnam, Dem. Rep.	32	32	72	117	126	131	131
WORLD TOTAL	89,209	91,010	93,602	96,514	98,430	100,221	101,894

Source: World Bank/FAO/UNID
Fertilizer Working Group
December 1986

ANNEX 3. Regional and World Non-Ammonia Nitrogen Supply ('000 Tons N)

	1984/85	1985/86	1986/87	1987/88	1988/89	1989/90	1990/91
DEVELOPED M.E.	320	320	320	320	320	320	320
North America	90	90	90	90	90	90	90
Western Europe	120	120	120	120	120	120	120
Oceania	5	5	5	5	5	5	5
Other Developed Ec.	105	105	105	105	105	105	105
DEVELOPING M.E.	45	45	45	45	45	45	50
Africa	0	0	0	0	0	0	0
Latin America	5	5	5	5	5	5	5
Near East	0	0	0	0	0	0	5
Far East	40	40	40	40	40	40	40
CENTRALLY PLANNED EC.	195	195	195	195	195	195	195
Eastern Europe	180	180	180	180	180	180	180
Socialist Asia	15	15	15	15	15	15	15
WORLD	560	560	560	560	560	560	565

ANNEX 4. FERTILIZER NITROGEN CONSUMPTION 1975-1984 ('000 Tons N)

	75/76	76/77	77/78	78/79	79/80	80/81	81/82	82/83	83/84	84/85
WORLD TOTAL	43,557.1	45,158.6	48,922.0	53,722.4	57,165.3	59,754.0	60,327.8	60,916.5	67,123.7	70,497.8
DEVELOPED M.E.	19,220.8	19,978.8	19,844.8	21,308.5	22,684.0	21,882.9	22,257.4	20,967.9	22,961.8	23,752.0
NORTH AMERICA	10,001.3	10,247.5	9,742.8	10,554.8	11,178.9	11,754.0	11,015.6	9,359.6	11,219.8	11,691.5
Canada	556.2	599.0	705.8	834.6	831.0	937.0	965.0	1,018.0	1,157.2	1,255.1
USA	9,445.1	9,648.5	9,037.0	9,720.2	10,347.9	10,817.0	10,050.6	8,341.6	10,062.6	10,436.4
WESTERN EUROPE	8,066.5	8,449.0	8,764.6	9,374.3	10,033.1	8,740.7	9,759.7	10,138.6	10,325.7	10,545.9
Austria	121.4	143.3	134.5	154.0	158.0	159.7	162.0	146.3	152.5	161.1
Belgium-Lux.	182.3	189.0	191.8	196.7	197.8	194.3	197.0	197.0	198.8	195.0
Denmark	339.1	349.5	373.7	379.9	393.9	374.1	376.0	391.4	419.0	408.4
Finland	199.3	161.4	168.4	183.3	196.2	197.0	182.4	216.0	204.8	196.1
France	1,707.8	1,815.0	1,831.7	1,978.0	2,134.8	2,146.5	2,193.0	2,196.4	2,320.0	2,336.8
Germany, F.R.	1,228.5	1,323.0	1,324.7	1,354.0	1,477.5	1,550.8	1,323.0	1,464.5	1,377.9	1,451.7
Greece	275.1	291.3	294.0	341.2	356.1	333.3	335.1	408.0	417.9	426.6
Iceland	14.8	14.8	15.0	15.8	15.7	14.9	15.2	14.0	14.8	13.5
Ireland	152.7	168.2	230.5	263.6	247.5	275.1	275.2	296.0	331.4	329.7
Italy	724.3	699.7	817.3	1,042.7	1,106.8	1,012.1	981.4	967.8	996.1	1,017.4
Malta	0.2	0.2	0.6	0.6	0.5	1.0	0.3	0.2	0.5	0.5
Netherlands	452.7	429.9	446.7	443.3	486.1	482.8	477.3	456.7	478.3	484.9
Norway	97.8	95.0	101.2	108.5	113.2	110.1	106.7	114.1	110.3	113.3
Portugal	141.0	136.6	154.4	138.6	155.0	136.6	142.8	138.8	125.7	123.2
Spain	722.2	830.4	788.1	807.6	908.2	901.9	818.3	798.9	808.5	914.8
Sweden	258.0	259.1	258.0	256.0	256.3	243.9	248.1	249.3	257.8	253.2
Switzerland	44.3	50.6	55.0	63.5	64.5	65.6	62.9	60.2	67.4	70.7
United Kingdom	1,045.0	1,110.0	1,177.0	1,222.0	1,314.0	124.0	1,386.0	1,560.0	1,601.0	1,580.0
Yugoslavia	360.0	382.0	402.0	425.0	440.0	417.0	477.0	463.0	443.0	469.0
OCEANIA	177.3	227.9	240.2	239.9	267.3	268.3	271.7	269.0	300.4	360.0
Australia	167.0	207.8	220.0	215.0	245.0	248.0	250.0	248.0	269.0	320.0
New Zealand	10.3	20.1	20.2	24.9	22.3	20.3	21.7	21.0	31.4	40.0
OTHER DEVELOPED M.E.	975.7	1,054.4	1,097.2	1,139.5	1,215.7	1,119.9	1,210.4	1,200.7	1,115.9	1,154.6
Israel	37.4	36.5	38.2	37.6	35.7	39.5	40.2	39.2	47.1	51.5
Japan	653.0	702.0	716.0	723.0	777.0	614.0	643.0	687.0	701.0	697.0
South Africa	285.3	315.9	343.0	378.9	403.0	466.4	527.2	474.5	367.8	406.1

FERTILIZER NITROGEN CONSUMPTION 1975-1984 ('000 Tons N) (Continued)

	75/76	76/77	77/78	78/79	79/80	80/81	81/82	82/83	83/84	84/85
DEVELOPING M.E.	7,323.2	8,463.0	9,560.0	10,140.4	11,170.1	12,307.8	12,762.2	13,213.7	14,410.1	15,809.4
AFRICA	410.1	445.6	457.9	484.4	535.9	681.0	657.0	626.2	670.8	694.9
Algeria	40.0	51.0	52.1	57.6	60.3	84.0	71.0	48.7	59.3	83.0
Angola	2.1	1.0	13.1	10.0	9.4	9.2	4.6	2.8	3.8	2.5
Benin	0.8	1.0	0.7	1.6	0.9	0.4	0.5	1.1	3.0	4.5
Botswana	1.5	1.0	1.0	1.0	0.5	0.6	0.6	0.6	0.5	0.5
Burkino Faso	0.4	2.2	1.8	3.4	2.0	1.2	1.6	3.7	4.1	3.5
Burundi	3.3	4.7	4.4	0.4	0.5	0.9	0.3	0.5	1.2	0.6
Cameroon	6.1	4.0	14.6	15.4	14.2	18.1	22.7	24.2	22.0	26.0
Central African R.	0.6	1.2	1.2	1.2	0.1	1.4	1.2	0.7	0.3	1.1
Chad	2.8	3.1	3.4	3.6	0.0	0.5	1.6	1.7	2.5	3.0
Congo	0.2	0.1	0.0	0.0	0.0	0.2	0.2	0.9	0.8	0.9
Cote d'Ivoire	10.9	12.0	17.5	15.9	16.6	19.7	11.3	7.4	10.0	10.0
Equatorial Guinea	0.1	0.1	0.1	0.1	0.1	0.1	0.0	0.0	0.0	0.2
Ethiopia	0.2	0.2	1.8	9.9	16.0	16.0	16.0	15.0	18.0	13.0
Gabon	0.3	0.4	0.1	0.1	0.1	0.1	0.8	0.2	0.6	0.5
Gambia	0.2	0.2	1.8	1.4	1.3	0.9	2.4	0.8	1.1	1.0
Ghana	11.0	12.0	10.1	10.6	8.4	5.0	15.0	12.5	11.0	5.0
Guinea	0.6	0.6	0.2	0.2	1.1	0.1	0.5	0.7	0.3	0.0
Kenya	21.9	22.4	25.3	25.4	20.1	26.7	33.9	34.2	31.0	36.0
Lesotho	0.3	0.3	0.3	0.4	0.4	0.5	0.5	0.5	0.5	0.6
Liberia	3.1	4.3	4.0	2.0	4.2	2.1	0.3	0.7	0.9	0.4
Madagascar	3.5	3.9	4.2	4.7	3.4	4.0	3.5	6.6	7.4	2.0
Malawi	12.9	18.9	12.3	19.6	14.4	22.8	14.3	24.2	23.1	28.1
Mali	1.6	4.5	7.9	7.3	4.0	7.8	5.7	2.4	8.6	15.0
Mauritania	1.1	0.5	1.9	1.9	1.2	0.4	0.0	0.0	0.4	0.4
Mauritius	10.0	8.8	12.0	9.1	9.8	10.2	9.6	9.9	10.7	10.7
Morocco	63.2	82.2	75.2	80.7	100.8	122.6	81.1	92.4	111.5	104.3
Mozambique	3.7	6.7	10.1	14.0	10.8	16.2	20.1	20.0	7.5	3.0
Niger	0.3	0.6	1.0	0.7	0.7	1.2	1.7	1.3	1.4	1.3
Nigeria	30.2	35.0	35.0	35.0	62.4	92.2	96.7	102.0	123.5	131.0
Reunion	4.8	4.1	5.7	6.1	5.1	1.8	5.3	3.5	6.3	2.8
Rwanda	0.1	0.2	0.1	0.1	0.1	0.1	0.1	0.4	0.5	1.0
Senegal	10.0	10.8	12.0	10.3	8.0	6.8	6.3	3.4	8.0	15.0
Sierra Leone	1.3	0.5	0.7	0.6	1.2	0.8	1.3	0.2	0.4	0.5
Somalia	1.8	0.0	0.1	0.1	0.1	1.2	1.3	0.8	1.8	2.6
Swaziland	5.5	1.0	1.4	1.6	3.7	7.5	8.5	8.5	10.0	4.5
Tanzania	14.9	13.3	16.0	12.9	23.0	22.8	17.5	16.4	15.5	23.5
Togo	0.8	0.9	0.3	0.8	1.6	0.9	1.2	1.1	1.2	2.1
Tunisia	23.4	24.9	17.2	23.5	27.0	23.8	30.5	39.9	32.8	41.5
Uganda	0.9	1.0	0.3	0.0	0.0	0.7	0.5	0.0	0.0	0.4
Zaire	5.0	6.2	4.5	4.7	5.1	3.7	2.5	2.1	4.1	4.5
Zambia	33.7	39.8	41.5	35.2	38.2	53.0	57.6	68.3	43.9	36.8
Zimbabwe	75.0	60.0	45.0	55.3	59.1	92.8	106.7	73.1	81.3	71.6

FERTILIZER NITROGEN CONSUMPTION 1975-1984 ('000 Tons N) (Continued)

	75/76	76/77	77/78	78/79	79/80	80/81	81/82	82/83	83/84	84/85
LATIN AMERICA	1,882.0	2,260.5	2,514.2	2,476.0	2,660.4	2,883.9	2,864.5	2,886.7	2,682.3	3,209.2
Argentina	28.0	39.2	41.7	47.7	58.9	62.6	51.2	53.2	64.6	98.0
Barbados	1.2	1.9	1.5	1.9	1.7	1.8	2.0	2.0	2.0	2.0
Bolivia	1.6	1.3	1.9	1.5	1.6	1.5	2.7	1.3	3.1	2.5
Brazil	406.2	498.3	700.5	705.9	783.4	926.1	667.7	642.3	553.1	813.8
Chile	36.7	48.6	37.4	50.0	50.7	51.4	48.8	48.3	64.9	86.3
Colombia	111.3	142.7	155.2	133.7	151.0	152.0	135.3	155.1	157.4	180.9
Costa Rica	31.1	28.5	40.5	40.9	39.4	40.5	45.2	49.0	54.0	50.6
Cuba	156.3	187.2	223.0	227.0	280.4	267.4	315.0	275.2	248.7	294.4
Dominican Republic	30.0	35.0	32.0	32.0	34.3	28.2	31.4	30.7	25.0	31.3
Ecuador	13.5	49.5	58.3	39.4	43.3	40.7	34.8	33.2	47.0	43.0
El Salvador	65.0	77.1	77.1	84.2	50.6	47.6	66.5	44.2	59.4	39.8
Guadeloupe	3.3	3.4	3.5	4.0	2.1	2.1	3.3	2.3	3.1	2.9
Guatemala	35.1	49.9	52.0	52.0	58.2	51.9	47.0	58.7	41.7	55.5
Guyana	7.7	6.0	6.3	12.6	6.5	4.3	9.3	7.2	9.6	10.0
Haiti	1.2	0.1	1.6	1.8	2.3	0.2	3.0	2.6	1.6	2.0
Honduras	11.4	11.0	15.2	9.9	10.4	16.0	16.7	12.8	19.0	23.0
Jamaica	6.0	7.3	4.5	6.4	6.4	8.9	10.2	6.0	6.7	8.8
Martinique	3.0	3.6	3.0	4.7	3.6	5.4	4.3	4.4	5.4	3.0
Mexico	732.2	830.4	794.2	752.2	826.1	904.3	1,111.7	1,254.6	1,087.8	1,193.1
Nicaragua	16.5	22.9	36.0	35.0	17.8	30.0	39.0	14.2	51.4	38.0
Panama	13.4	11.0	9.0	9.6	11.9	11.4	13.9	13.9	11.0	12.2
Paraguay	0.3	0.5	0.3	0.8	1.1	1.3	2.3	2.0	1.7	1.3
Peru	83.6	100.3	107.1	103.9	88.4	84.8	100.6	71.2	52.9	53.8
St. Kitts, Nevis, Ang.	0.3	0.3	0.3	0.3	0.3	0.3	0.3	0.4	0.4	0.5
St. Lucia	1.6	1.6	2.0	1.1	1.4	0.7	0.8	0.8	1.4	1.4
St. Vincent	2.1	2.1	0.4	2.4	2.4	2.4	2.4	2.4	2.4	2.5
Surinam	2.5	3.4	3.7	3.5	1.9	1.2	5.4	7.1	12.0	10.0
Trinidad & Tobago	4.3	4.5	4.6	4.9	4.7	3.9	2.8	3.4	3.9	5.0
Uruguay	11.1	17.9	17.8	14.4	21.8	21.1	20.4	16.5	16.7	19.5
Venezuela	64.8	74.3	82.8	91.5	97.0	113.0	69.6	70.8	73.4	123.1
Virgin Islands	0.7	0.7	0.8	0.8	0.8	0.9	0.9	0.9	1.0	1.0

FERTILIZER NITROGEN CONSUMPTION 1975-1984 ('000 Tons N) (Continued)

	75/76	76/77	77/78	78/79	79/80	80/81	81/82	82/83	83/84	84/85
NEAR EAST	1,319.2	1,491.9	1,562.6	1,650.7	1,725.3	1,993.8	2,131.4	2,411.1	2,786.7	2,610.5
Afghanistan	27.7	30.6	37.0	48.7	33.2	37.8	25.6	31.4	36.8	40.0
Cyprus	15.0	15.0	8.2	9.9	9.1	8.2	9.0	11.6	10.6	10.4
Egypt	415.0	427.7	459.5	490.5	500.0	554.0	585.0	667.8	722.2	649.2
Iran	190.0	220.5	221.1	167.7	237.0	274.8	386.6	492.8	596.6	486.6
Iraq	25.0	35.0	45.0	43.4	74.2	64.5	63.6	62.1	61.3	73.0
Jordan	1.9	4.0	1.4	3.2	6.2	6.0	2.9	7.1	8.8	6.9
Kuwait	0.0	0.0	0.1	0.1	0.3	0.1	0.3	0.6	0.8	0.5
Lebanon	7.0	17.7	13.7	14.4	16.0	14.0	16.9	17.3	18.0	17.4
Libyan Arab Rep.	15.0	13.1	16.7	13.8	22.5	29.5	30.5	35.2	33.9	56.4
Qatar	0.2	0.3	0.3	0.4	0.6	0.8	0.8	0.7	0.6	0.6
Saudi Arabia	5.7	6.4	4.7	7.0	16.0	25.2	41.5	59.0	113.3	130.0
Sudan	95.0	80.0	32.0	11.2	33.8	80.4	74.5	54.0	36.8	41.4
Syrian Arab Rep.	35.2	48.1	55.2	60.9	78.1	79.8	83.1	95.9	109.5	126.7
Turkey	482.2	590.7	665.7	773.8	685.5	807.5	798.8	863.4	1,021.4	954.8
Yemen Arab Rep.	3.4	2.0	1.9	4.7	11.7	9.4	10.6	10.4	14.1	14.6
Yeman, P.D.R.	0.9	0.8	0.1	1.0	1.1	1.8	1.7	1.8	2.0	2.0
FAR EAST	3,711.9	4,265.0	5,025.3	5,529.3	6,248.5	6,749.1	7,109.3	7,289.7	8,270.3	9,294.8
Bangladesh	146.7	165.8	223.2	227.7	260.2	267.9	251.6	306.0	356.4	399.0
Burma	42.5	42.9	49.9	67.9	69.7	66.8	92.8	114.7	115.0	127.4
Fiji	7.2	9.0	9.9	9.0	12.1	10.8	12.8	10.3	7.5	10.1
India	1,908.7	2,351.7	2,813.4	2,986.3	3,444.2	3,522.0	3,881.0	4,043.0	4,627.3	5,333.3
Indonesia	341.9	351.2	465.2	524.0	620.4	950.9	997.0	891.0	1,049.1	1,285.4
Korea, Rep. of	467.7	361.3	387.9	461.6	443.9	447.2	431.8	308.6	367.3	401.8
Laos	0.1	0.1	0.1	0.0	0.1	4.0	4.0	0.5	0.5	0.0
Malaysia	77.6	90.0	102.0	109.5	137.7	139.3	127.9	138.0	235.0	249.0
Nepal	8.4	10.7	13.0	14.0	15.5	16.4	17.3	23.0	33.0	31.6
Pakistan	443.5	511.0	554.1	684.1	775.8	843.6	832.6	958.6	914.3	934.8
Papua New Guinea	3.4	4.4	5.8	6.1	3.2	3.0	7.3	3.2	3.5	3.4
Philippines	144.1	177.2	174.3	205.4	226.7	224.8	210.7	231.4	240.2	177.9
Singapore	1.0	1.0	1.0	1.0	1.8	1.8	2.0	2.0	2.0	2.0
Sri Lanka	40.1	52.3	65.4	79.2	77.2	91.6	78.5	79.4	83.0	96.1
Thailand	78.9	136.3	160.1	153.5	160.0	159.0	162.0	180.0	236.2	243.0

FERTILIZER NITROGEN CONSUMPTION

	75/76	76/77	77/78	78/79	79/80	80/81	81/82	82/83	83/84	84/85
CENT. PLANNED EC.	17,013.2	16,716.9	19,517.2	22,273.5	23,311.2	25,563.3	25,308.2	26,734.9	29,751.8	30,936.4
ASIA	5,498.5	5,386.7	7,807.5	10,031.1	11,306.6	12,829.9	12,311.8	13,068.6	14,626.0	15,995.3
China a/	5,039.0	4,755.0	7,073.0	9,280.0	10,641.0	12,111.0	11,528.0	12,210.0	13,678.9	15,076.0
Korea, D.P.R.	264.0	385.8	442.8	535.0	540.0	550.0	564.1	591.6	582.8	597.0
Viet Nam, Dem. Rep.	194.0	241.5	288.0	209.0	121.0	156.0	199.7	250.0	351.0	310.0
Mongolia	1.5	4.4	3.7	7.1	4.6	5.7	9.5	9.7	11.7	11.8
Kampuchea D.P.						7.2	10.5	7.3	1.6	0.5
EASTERN EUROPE	11,514.7	11,330.2	11,709.7	12,242.4	12,004.6	12,733.4	12,996.4	13,666.3	15,125.8	14,941.1
Albania	36.0	45.3	50.0	55.0	70.9	67.7	72.5	73.0	73.0	75.0
Bulgaria	378.0	348.8	362.6	376.3	420.5	450.0	518.0	536.0	550.0	479.0
Czechoslvakia	532.0	582.0	601.0	628.0	630.0	675.0	615.0	646.0	682.0	691.7
Germany, Dem. Rep.	678.0	754.9	770.8	784.0	747.6	751.8	750.0	607.0	693.9	697.0
Hungary	535.8	522.4	582.0	570.1	570.1	536.8	562.9	647.0	625.4	625.9
Poland	1,227.9	1,212.8	1,250.3	1,350.0	1,312.5	1,343.8	1,213.0	1,239.3	1,321.5	1,238.5
Romania	788.0	612.0	571.0	821.0	786.0	646.3	882.0	880.0	878.0	857.0
USSR	7,339.0	7,252.0	7,522.0	7,658.0	7,467.0	8,262.0	8,383.0	9,038.0	10,302.0	10,277.0

a/ Includes Taiwan.

Source: FAO Handbook

ANNEX 5. Regional and World Nitrogen Fertilizer Demand 1985-1995

	1984/85	1985/86	1986/87	1987/88	1988/89	1989/90	1990/91	1995/96
DEVELOPED M.E.	23,750	23,150	22,710	23,050	23,370	24,380	24,630	25,960
North America	11,690	10,900	10,290	10,440	10,610	11,500	11,630	12,500
Western Europe	10,540	10,700	10,860	11,020	11,150	11,250	11,350	11,700
Oceania	360	370	370	380	390	400	410	460
Other Developed M.E.	1,160	1,180	1,190	1,210	1,220	1,230	1,240	1,300
DEVELOPING M.E.	15,810	16,510	17,570	18,650	19,810	21,010	22,210	27,470
Africa	690	830	870	910	950	990	1,030	1,220
Latin America	3,210	3,150	3,420	3,690	3,990	4,290	4,590	5,600
Near East	2,610	2,590	2,740	2,880	3,030	3,180	3,290	3,850
Far East	9,300	9,940	10,540	11,170	11,840	12,550	13,300	16,800
CENTRALLY PLANNED EC.	30,940	30,500	31,550	32,650	33,800	35,000	36,200	39,900
Eastern Europe	14,940	15,500	16,150	16,800	17,500	18,200	18,900	21,400
Socialist Asia	16,000	15,000	15,400	15,850	16,300	16,800	17,300	18,500
WORLD	70,500	70,160	71,830	74,350	76,980	80,390	83,040	93,330

ANNEX 6. Regional and World Industrial and Feed Nitrogen Demand ('000 Tons N)

	1984/85	1985/86	1986/87	1987/88	1988/89	1989/90	1990/91
DEVELOPED M.E.	7,040	7,076	7,163	7,242	7,328	7,415	7,507
North America	2,982	3,012	3,042	3,072	3,103	3,134	3,165
Western Europe	2,538	2,510	2,529	2,543	2,560	2,582	2,604
Oceania	118	121	124	128	131	134	138
Other Developed M.E.	1,402	1,433	1,468	1,499	1,534	1,565	1,600
DEVELOPING M.E.	658	689	715	742	766	795	820
Africa	49	50	51	52	54	55	56
Latin America	251	263	275	288	298	312	323
Near East	70	77	80	82	84	86	88
Far East	288	299	309	320	330	342	353
CENTRALLY PLANNED EC.	2,640	2,701	2,756	2,815	2,869	2,931	2,994
Eastern Europe	2,040	2,086	2,131	2,175	2,219	2,266	2,314
Socialist Asia	600	615	625	640	650	665	680
WORLD	10,338	10,466	10,634	10,799	10,963	11,141	11,321

Regional and World Industrial and Feed Nitrogen Demand ('000 Tons N) (Continued)

	1984/85	1985/86	1986/87	1987/88	1988/89	1989/90	1990/91
DEVELOPED M.E.	7,040	7,076	7,163	7,242	7,328	7,415	7,507
North America	2,982	3,012	3,042	3,072	3,103	3,134	3,165
Canada	282	290	300	310	320	330	340
U.S.A.	2,700	2,722	2,742	2,762	2,783	2,804	2,825
Western Europe	2,538	2,510	2,529	2,543	2,560	2,582	2,604
Austria	94	94	96	96	97	97	97
Belgium-Lux.	49	49	51	51	51	51	51
Denmark	18	20	20	21	21	22	22
Finland	25	25	25	25	25	25	25
France	433	462	471	480	490	500	510
Germany F.R.	653	606	604	605	593	592	592
Greece	10	11	11	12	12	13	13
Ireland	4	4	4	4	4	4	4
Italy	359	360	360	362	364	364	364
Netherlands	267	252	257	262	267	272	277
Norway	36	36	36	36	36	36	36
Portugal	21	21	22	22	22	22	22
Spain	101	103	110	116	122	128	134
Sweden	60	59	59	59	59	59	59
Switzerland	19	19	19	19	19	19	19
United Kingdom	370	369	363	351	355	354	354
Yugoslavia	19	20	21	22	23	24	25
Oceania	118	121	124	128	131	134	138
Australia	113	116	119	122	125	128	131
New Zealand	5	5	5	6	6	6	7
Other Developed M.E.	1,402	1,433	1,468	1,499	1,534	1,565	1,600
Israel	12	13	13	14	14	15	15
Japan	1,190	1,220	1,250	1,280	1,310	1,340	1,370
South Africa	200	200	205	205	210	210	215
DEVELOPING M.E.	658	689	715	742	766	795	820
Africa	49	50	51	52	54	55	56
Algeria	13	14	15	16	17	18	19
Zambia	11	11	11	11	12	12	12
Zimbabwe	25	25	25	25	25	25	25

(Table continued on the following page)

Regional and World Industrial and Feed Nitrogen Demand ('000 Tons N) (Continued)

	1984/85	1985/86	1986/87	1987/88	1988/89	1989/90	1990/91
Latin America	251	263	275	288	298	312	323
Argentina	14	14	15	15	15	16	16
Brazil	95	100	105	110	115	120	125
Colombia	17	18	18	19	19	20	20
Cuba	11	12	12	13	13	14	14
Mexico	95	100	105	110	115	120	125
Peru	9	9	10	10	10	11	11
Venezuela	10	10	10	11	11	11	12
Near East	70	77	80	82	84	86	88
Egypt	35	36	37	38	39	40	41
Iran	0	5	5	5	5	5	5
Turkey	35	36	38	39	40	41	42
Far East	288	299	309	320	330	342	353
India	68	71	74	77	80	83	86
Indonesia	99	103	107	111	115	120	125
Korea Rep. of	77	79	81	83	85	87	89
Malaysia	33	34	35	36	37	38	39
Philippines	11	12	12	13	13	14	14
CENTRALLY PLANNED EC.	2,640	2,701	2,756	2,815	2,869	2,931	2,994
Eastern Europe	2,040	2,086	2,131	2,175	2,219	2,266	2,314
Bulgaria	66	69	72	75	78	81	84
Czechoslovakia	92	95	98	100	102	105	108
Germany, Dem. Rep.	230	240	250	260	270	280	290
Hungary	93	96	99	102	105	108	111
Poland	92	95	98	100	102	105	108
Romania	77	80	82	85	87	90	93
U.S.S.R.	1,390	1,411	1,432	1,453	1,475	1,497	1,520
Socialist Asia	600	615	625	640	650	665	680
China	470	480	490	500	510	520	530
Taiwan	130	135	135	140	140	145	150
WORLD TOTAL	10,338	10,466	10,634	10,799	10,963	11,141	11,321

Source: World Bank/FAO/UNIDO
Fertilizer Working Group
December 19866

ANNEX 7. Regional and World Nitrogen Supply-Demand Balances ('000 Tons N)

	1984/85	1985/86	1986/87	1987/88	1988/89	1989/90	1990/91
DEVELOPED M.E.							
North America							
NH3 Nominal Capacity	16,504	16,794	17,183	17,508	17,508	17,508	17,508
NH3 Supply Capability	15,158	14,994	15,228	15,983	16,165	16,208	16,222
NH3 Industrial Use	2,982	3,012	3,042	3,072	3,103	3,134	3,165
Losses	852	839	853	904	914	915	914
NH3 Available for Fert.	11,324	11,143	11,333	12,007	12,148	12,159	12,143
Non-NH3 Nitrogen	90	90	90	90	90	90	90
N Fert. Supply Potential	11,414	11,233	11,423	12,097	12,238	12,249	12,233
N Fert. Consumption	11,690	10,900	10,290	10,440	10,610	11,500	11,630
Surplus (-Deficit)	(276)	333	1,133	1,657	1,628	749	603
Western Europe							
NH3 Nominal Capacity	15,299	15,466	15,421	15,271	15,502	15,867	15,867
NH3 Supply Capability	13,465	13,506	13,874	13,861	13,884	14,093	14,288
NH3 Industrial Use	2,538	2,510	2,529	2,543	2,560	2,582	2,604
Losses	765	770	794	792	793	806	818
NH3 Available for Fert.	10,162	10,226	10,551	10,526	10,531	10,705	10,866
Non-NH3 Nitrogen	120	120	120	120	120	120	120
N Fert. Supply Potential	10,282	10,346	10,671	10,646	10,651	10,825	10,986
N Fert. Consumption	10,540	10,700	10,860	11,020	11,150	11,250	11,350
Surplus (-Deficit)	(258)	(354)	(189)	(374)	(499)	(425)	(364)
Oceania							
NH3 Nominal Capacity	525	525	525	525	525	525	890
NH3 Supply Capability	473	476	476	476	476	476	476
NH3 Industrial Use	118	121	124	128	131	134	138
Losses	25	25	25	24	24	24	24
NH3 Available for Fert.	330	330	327	324	321	318	314
Non-NH3 Nitrogen	5	5	5	5	5	5	5
N Fert. Supply Potential	335	335	332	329	326	323	319
N Fert. Consumption	360	370	370	380	390	400	410
Surplus (-Deficit)	(25)	(35)	(38)	(51)	(64)	(77)	(91)
Other Developed M.E.							
NH3 Nominal Capacity	3,018	2,480	2,480	2,480	2,480	2,480	2,480
NH3 Supply Capability	2,733	2,433	2,220	2,232	2,232	2,232	2,232
NH3 Industrial Use	1,402	1,433	1,468	1,499	1,534	1,565	1,600
Losses	93	70	53	51	49	47	44
NH3 Available for Fert.	1,238	930	699	682	649	620	588
Non-NH3 Nitrogen	105	105	105	105	105	105	105
N Fert. Supply Potential	1,343	1,035	804	787	754	725	693
N Fert. Consumption	1,160	1,180	1,190	1,210	1,220	1,230	1,240
Surplus (-Deficit)	183	(145)	(386)	(423)	(466)	(505)	(547)

(Table continued on the following page)

Regional and World Nitrogen Supply-Demand Balances ('000 Tons N) (Continued)

	1984/85	1985/86	1986/87	1987/88	1988/89	1989/90	1990/91
DEVELOPING M.E.							
Africa							
NH3 Nominal Capacity	404	747	1,019	1,019	1,291	1,291	1,291
NH3 Supply Capability	233	307	307	523	632	737	759
NH3 Industrial Use	49	50	51	52	54	55	56
Losses	18	26	26	47	58	68	70
NH3 Available for Fert.	166	231	230	424	520	614	633
Non-NH3 Nitrogen	0	0	0	0	0	0	0
N Fert. Supply Potential	166	231	230	424	520	614	633
N Fert. Consumption	690	830	870	910	950	990	1,030
Surplus (-Deficit)	(524)	(599)	(640)	(486)	(430)	(376)	(397)
Latin America							
NH3 Nominal Capacity	5,833	5,833	5,833	5,833	6,203	6,620	7,235
NH3 Supply Capability	4,785	4,838	4,850	4,850	4,998	5,300	5,699
NH3 Industrial Use	251	263	275	288	298	312	323
Losses	453	458	458	456	470	499	538
NH3 Available for Fert.	4,081	4,118	4,118	4,106	4,230	4,489	4,838
Non-NH3 Nitrogen	5	5	5	5	5	5	5
N Fert. Supply Potential	4,086	4,123	4,123	4,111	4,235	4,494	4,843
N Fert. Consumption	3,210	3,150	3,420	3,690	3,990	4,290	4,590
Surplus (-Deficit)	876	973	703	421	245	204	253
Near East							
NH3 Nominal Capacity	4,286	4,884	4,884	5,374	5,646	5,918	6,299
NH3 Supply Capability	3,135	3,399	3,931	4,160	4,457	4,646	4,864
NH3 Industrial Use	70	77	80	82	84	86	88
Losses	307	332	385	408	437	456	478
NH3 Available for Fert.	2,759	2,990	3,466	3,670	3,936	4,104	4,298
Non-NH3 Nitrogen	0	0	0	0	0	0	5
N Fert. Supply Potential	2,759	2,990	3,466	3,670	3,936	4,104	4,303
N Fert. Consumption	2,610	2,590	2,740	2,880	3,030	3,180	3,290
Surplus (-Deficit)	149	400	726	790	906	924	1,013
Far East							
NH3 Nominal Capacity	11,090	12,424	12,750	13,217	14,219	15,493	16,223
NH3 Supply Capability	8,087	8,838	9,527	10,346	10,899	11,706	12,478
NH3 Industrial Use	288	299	309	320	330	342	353
Losses	780	854	922	1,003	1,057	1,136	1,213
NH3 Available for Fert.	7,019	7,685	8,296	9,023	9,512	10,228	10,913
Non-NH3 Nitrogen	40	40	40	40	40	40	40
N Fert. Supply Potential	7,059	7,725	8,336	9,063	9,552	10,268	10,953
N Fert. Consumption	9,300	9,940	10,540	11,170	11,840	12,550	13,300
Surplus (-Deficit)	(2,241)	(2,215)	(2,204)	(2,107)	(2,288)	(2,282)	(2,348)

(Table continued on the following page

Regional and World Nitrogen Supply-Demand Balances ('000 Tons N) (Continued)

	1984/85	1985/86	1986/87	1987/88	1988/89	1989/90	1990/91
CENT.PLANNED EC.							
Eastern Europe							
NH3 Nominal Capacity	33,467	34,419	34,789	34,993	34,993	34,993	34,993
NH3 Supply Capability	26,135	27,081	27,701	28,034	28,181	28,211	28,219
NH3 Industrial Use	2,040	2,086	2,131	2,175	2,219	2,266	2,314
Losses	2,410	2,500	2,557	2,586	2,596	2,595	2,591
NH3 Available for Fert.	21,686	22,496	23,013	23,273	23,366	23,351	23,315
Non-NH3 Nitrogen	180	180	180	180	180	180	180
N Fert. Supply Potential	21,866	22,676	23,193	23,453	23,546	23,531	23,495
N Fert. Consumption	14,940	15,500	16,150	16,800	17,500	18,200	18,900
Surplus (-Deficit)	6,926	7,176	7,043	6,653	6,046	5,331	4,595
Socialist Asia							
NH3 Nominal Capacity	17,608	18,152	18,617	19,661	19,661	19,661	19,661
NH3 Supply Capability	15,005	15,138	15,488	16,049	16,506	16,612	16,657
NH3 Industrial Use	600	615	625	640	650	665	680
Losses	1,729	1,743	1,784	1,849	1,903	1,914	1,917
NH3 Available for Fert.	12,676	12,780	13,079	13,560	13,953	14,033	14,060
Non-NH3 Nitrogen	15	15	15	15	15	15	15
N Fert. Supply Potential	12,691	12,795	13,094	13,575	13,968	14,048	14,075
N Fert. Consumption	16,000	15,000	15,400	15,850	16,300	16,800	17,300
Surplus (-Deficit)	(3,309)	(2,205)	(2,306)	(2,275)	(2,332)	(2,752)	(3,225)

(Table continued on the following page)

Regional and World Nitrogen Supply-Demand Balances ('000 Tons N) (Continued)

	1984/85	1985/86	1986/87	1987/88	1988/89	1989/90	1990/91
ALL DEVELOPED M.E.							
NH3 Nominal Capacity	35,346	35,265	35,609	35,784	36,015	36,380	36,745
NH3 Supply Capability	31,829	31,409	31,798	32,552	32,757	33,009	33,218
NH3 Industrial Use	7,040	7,076	7,163	7,242	7,328	7,415	7,507
Losses	1,735	1,703	1,724	1,772	1,780	1,792	1,800
NH3 Available for Fert.	23,054	22,630	22,911	23,538	23,649	23,802	23,911
Non-NH3 Nitrogen	320	320	320	320	320	320	320
N Fert. Supply Potential	23,374	22,950	23,231	23,858	23,969	24,122	24,231
N Fert. Consumption	23,750	23,150	22,710	23,050	23,370	24,380	24,630
Surplus (-Deficit)	(376)	(200)	521	808	599	(258)	(399)
ALL DEVELOPING M.E.							
NH3 Nominal Capacity	21,613	23,888	24,486	25,443	27,359	29,322	31,048
NH3 Supply Capability	16,240	17,382	18,615	19,879	20,986	22,389	23,800
NH3 Industrial Use	658	689	715	742	766	795	820
Losses	1,558	1,669	1,790	1,914	2,022	2,159	2,298
NH3 Available for Fert.	14,024	15,024	16,110	17,223	18,198	19,435	20,682
Non-NH3 Nitrogen	45	45	45	45	45	45	50
N Fert. Supply Potential	14,069	15,069	16,155	17,268	18,243	19,480	20,732
N Fert. Consumption	15,810	16,510	17,570	18,650	19,810	21,010	22,210
Surplus (-Deficit)	(1,741)	(1,441)	(1,415)	(1,382)	(1,567)	(1,530)	(1,478)
ALL CENT. PLANNED EC.							
NH3 Nominal Capacity	51,075	52,571	53,406	54,654	54,654	54,654	54,654
NH3 Supply Capability	41,140	42,219	43,189	44,083	44,687	44,823	44,876
NH3 Industrial Use	2,640	2,701	2,756	2,815	2,869	2,931	2,994
Losses	4,138	4,242	4,341	4,435	4,499	4,508	4,508
NH3 Available for Fert.	34,362	35,276	36,092	36,833	37,319	37,384	37,374
Non-NH3 Nitrogen	195	195	195	195	195	195	195
N Fert. Supply Potential	34,557	35,471	36,287	37,028	37,514	37,579	37,569
N Fert. Consumption	30,940	30,500	31,550	32,650	33,800	35,000	36,200
Surplus (-Deficit)	3,617	4,971	4,737	4,378	3,714	2,579	1,369
WORLD TOTAL							
NH3 Nominal Capacity	108,034	111,724	113,501	115,881	118,028	120,356	122,447
NH3 Supply Capability	89,209	91,010	93,602	96,514	98,430	100,221	101,894
NH3 Industrial Use	10,338	10,466	10,634	10,799	10,963	11,141	11,321
Losses	7,432	7,615	7,855	8,120	8,301	8,459	8,606
NH3 Available for Fert.	71,439	72,929	75,113	77,595	79,166	80,621	81,967
Non-NH3 Nitrogen	560	560	560	560	560	560	565
N Fert. Supply Potential	71,999	73,489	75,673	78,155	79,726	81,181	82,532
N Fert. Consumption	70,500	70,160	71,830	74,350	76,980	80,390	83,040
Surplus (-Deficit)	1,499	3,329	3,843	3,805	2,746	791	(508)

Source: World Bank/FAO/UNIDO
Fertilizer Working Group
December 1986

ANNEX 8. MAJOR NITROGEN CONSUMING COUNTRIES (MILLION TONS N)

Year	World Total	Developed Economies				Developing Economies	Centrally Planned Economies		
		USA	France	UK	GFR	India	USSR	Poland	China a/
1964/65	15.24	4.21	0.86	0.60	0.78	0.54	1.76	0.39	NA
1965/66	17.28	4.83	0.87	0.69	0.87	0.54	2.28	0.43	NA
1966/67	22.09	5.46	0.99	0.76	0.89	0.84	2.66	0.51	NA
1967/68	24.36	6.16	1.13	0.91	0.95	1.05	3.09	0.60	1.96
1968/69	26.47	6.31	1.24	0.85	0.93	1.25	3.45	0.70	2.15
1969/70	28.36	6.77	1.24	0.69	1.08	1.04	3.80	0.79	2.63
1970/71	31.80	7.36	1.45	0.80	1.13	1.31	4.60	0.82	3.37
1971/72	33.39	7.26	1.52	0.93	1.13	1.65	5.18	0.91	3.37
1972/73	36.06	7.51	1.59	0.79	1.19	1.74	5.60	0.98	3.86
1973/74	38.99	8.30	1.83	0.87	1.10	1.61	6.22	1.07	4.45
1974/75	38.50	7.81	1.55	0.93	1.20	1.84	6.70	1.14	3.85
1975/76	43.58	9.44	1.71	1.04	1.23	1.91	7.34	1.23	5.04
1976/77	45.22	9.66	1.81	1.11	1.32	2.53	7.25	1.21	4.76
1977/78	49.01	9.04	1.83	1.18	1.32	2.81	7.52	1.25	7.07
1978/79	53.84	9.72	1.98	1.22	1.35	2.99	7.66	1.30	9.28
1979/80	57.17	10.35	2.13	1.31	1.48	3.44	7.47	1.31	10.64
1980/81	60.73	10.82	2.15	1.24	1.55	3.52	8.26	1.34	12.11
1981/82	60.30	9.96	2.19	1.39	1.32	3.88	8.38	1.21	11.53
1982/83	61.02	8.34	2.20	1.56	1.46	4.04	9.04	1.24	12.21
1983/84	67.13	10.06	2.32	1.60	1.38	4.63	10.30	1.32	13.68
1984/85	70.50	10.44	2.34	1.58	1.45	5.33	10.28	1.23	15.07

a/ Apparent Consumption including Taiwan.
Actual figures will be lower due to stock changes.

Source: FAO Fertilizer Handbook

ANNEX 9. NEW AMMONIA CAPACITY COMMISSIONED 1960-1990 ('000 TONS)

	1960/61	1961/62	1962/63	1963/64	1964/65	1965/66
North America	-	420	132	549	40	735
Western Europe	27	-	245	464	91	486
Oceania and Other Developed	-	-	91	82	-	74
Total Developed ME	27	420	468	1,095	131	1,295
Africa	-	-	-	-	-	-
Latin America	-	13	184	-	213	322
Near East	-	125	33	38	-	-
Far East	-	54	-	-	-	116
Total Developing ME	-	192	217	38	213	438
Eastern Europe	25	197	362	312	1,145	398
Socialist Asia	-	145	-	-	-	-
Total Cent. Planned Ec.	25	342	362	312	1,145	398
World Total	52	954	1,047	1,445	1,489	2,131

	1975/76	1976/77	1977/78	1978/79	1979/80	1980/81
North America	706	481	3,291	637	569	511
Western Europe	255	252	761	450	778	46
Oceania and Other Developed	41	313	174	-	-	-
Total Developed ME	1,002	1,046	4,226	1,087	1,347	557
Africa	-	-	-	-	-	-
Latin America	27	-	641	-	-	-
Near East	156	104	-	544	516	-
Far East	523	559	1,036	343	1,328	675
Total Developing ME	706	663	1,677	887	1,844	675
Eastern Europe	1,689	1,242	494	1,803	4,219	534
Socialist Asia	-	1,033	544	519	1,550	446
Total Cent. Planned Ec.	1,689	2,275	1,038	2,322	5,769	980
World Total	3,397	3,984	6,941	4,296	8,960	2,212

Industry Department
March 1987

NEW AMMONIA CAPACITY COMMISSIONED 1960-1990 - '000 TONS

1966/67	1967/68	1968/69	1969/70	1970/71	1971/72	1972/73	1973/74	1975/75
1,709	1,889	464	789	1,459	81	885	-	254
399	862	1,273	977	1,251	1,940	165	740	1,288
-	940	238	144	407	556	-	-	-
2,108	3,691	1,975	1,910	3,117	2,577	1,050	740	1,542
-	-	-	-	25	-	50	-	-
-	123	54	163	925	176	-	665	866
109	-	-	163	92	476	41	-	244
41	368	110	576	202	300	93	342	629
150	491	164	902	1,244	952	184	1,007	1,739
1,213	363	1,352	1,357	1,925	301	1,011	1,306	2,522
-	100	308	-	-	-	-	-	-
1,213	463	1,660	1,357	1,925	301	1,011	1,306	2,522
3,471	4,645	3,799	4,169	6,286	3,830	2,245	3,053	5,803

1981/82	1982/83	1983/84	1984/85	1985/86	1986/87	1987/88	1988/89	1989/90
59	159	989	-	442	389	325	-	-
-	384	370	836	357	-	-	779	365
82	-	388	41	-	-	-	-	-
141	543	1,747	877	799	389	325	779	365
-	329	-	46	343	272	-	272	-
-	597	246	-	-	-	-	370	417
597	272	544	300	598	-	490	272	272
655	1,077	286	1,141	1,481	326	516	1,002	1,274
1,252	2,275	1,076	1,487	2,422	598	1,006	1,916	1,963
740	2,590	2,229	944	1,052	370	204	-	-
-	-	272	-	544	465	1,044	-	-
740	2,590	2,501	944	1,596	835	1,248	-	-
2,133	5,408	5,324	3,308	4,817	1,822	2,579	2,695	2,328

ANNEX 10. WORLD AMMONIA TRADE 1984
(thousand tonnes N)

Source / Destination	France	Germany, F.R.	Netherlands	United Kingdom	Other Western Europe	USSR	Other Eastern Europe	USA	Mexico	Trinidad	Venezuela	Arab Gulf	Indonesia	Libya	Others	Total
Western Europe	191.9	288.4	524.3	166.8	260.8	1,240.4	161.4	38.1	108.5	286.0	75.3	8.4	-	63.3	42.0	3,455.6
Austria	-	-	-	-	-	7.4	44.8	-	-	-	-	-	-	-	-	55.2
Belgium	7.8	2.8	233.8	32.9	-	75.0	-	7.4	-	42.7	22.0	-	-	7.4	-	431.8
Denmark	-	175.0	6.6	13.2	23.7	53.4	-	-	-	72.3	-	-	-	-	6.0	350.2
Finland	-	9.5	68.1	22.8	9.6	179.3	-	-	-	-	7.7	-	-	-	-	297.0
France	-	27.2	16.2	13.2	29.3	6.6	-	-	9.9	47.3	-	-	-	-	-	149.7
Germany, F.R.	96.3	-	26.8	6.7	26.4	-	76.0	-	-	-	-	-	-	-	-	232.3
Greece	-	-	-	6.6	15.4	137.0	-	-	-	6.4	-	-	-	1.6	-	167.1
Iceland	-	-	-	-	5.2	-	-	-	-	-	-	-	-	-	-	5.2
Ireland	-	0.1	8.8	0.2	-	-	-	-	-	-	-	-	-	-	-	9.1
Italy	0.1	0.7	-	-	26.6	153.7	0.1	-	-	-	-	-	-	-	-	181.2
Netherlands	1.5	2.8	-	12.1	10.2	13.2	-	-	-	3.4	-	-	-	-	-	43.2
Norway	4.5	-	2.5	-	-	18.2	-	-	-	-	-	-	-	-	-	25.2
Portugal	3.2	-	6.0	7.0	-	-	4.9	-	-	-	-	-	-	9.9	1.7	32.7
Spain	49.6	18.8	50.1	45.7	59.9	72.7	25.0	9.0	86.3	32.2	16.3	-	-	24.7	32.7	523.0
Sweden	-	0.3	6.7	-	13.1	149.7	7.6	-	12.3	-	-	-	-	-	-	189.7
Switzerland	6.4	1.0	0.1	-	12.3	-	-	-	-	-	-	-	-	-	-	19.8
Turkey	14.1	-	5.8	6.6	16.9	315.2	-	21.7	-	50.2	19.4	-	-	19.7	1.6	471.2
United Kingdom	8.4	50.1	93.0	-	12.0	58.9	-	-	-	31.2	9.9	-	-	-	-	272.0
Eastern Europe	-	-	-	-	23.2	159.1	58.9	-	-	-	-	-	-	-	-	241.2
Czechoslovakia	-	-	-	-	-	41.1	-	-	-	-	-	-	-	-	-	41.1
Hungary	-	-	-	-	-	-	2.7	-	-	-	-	-	-	-	-	2.7
Poland	-	-	-	-	-	98.6	-	-	-	-	-	-	-	-	-	98.6
Yugoslavia	-	-	-	-	23.2	19.3	56.2	-	-	-	-	-	-	-	-	98.7
Africa	0.2	3.5	0.1	0.1	6.0	136.7	-	16.6	6.9	7.1	-	1.0	-	47.7	83.0	308.9
Algeria	-	-	-	-	-	11.0	-	-	5.0	-	-	-	-	-	-	16.0
Cote d'Ivoire	-	-	-	-	-	-	-	-	-	-	-	1.0	-	-	-	1.0
Liberia	-	-	-	-	-	-	-	0.5	-	-	-	-	-	-	-	0.5
Mauritius	-	-	-	-	-	10.4	-	-	-	-	-	-	-	-	-	10.4
Morocco	0.1	3.3	-	-	-	30.4	-	-	1.9	-	-	-	-	-	2.9	38.6
Mozambique	-	-	-	-	-	25.5	-	-	-	-	-	-	-	-	30.6	56.1
Senegal	-	-	-	-	5.8	-	-	-	-	7.1	-	-	-	-	-	12.9
South Africa	-	-	-	-	-	-	-	12.9	-	-	-	-	-	-	-	12.9
Tanzania	-	-	-	-	-	-	-	2.9	-	-	-	-	-	-	-	2.9
Tunisia	-	-	-	-	0.2	59.4	-	-	-	-	-	-	-	47.7	26.3	133.6
Zambia	-	-	-	-	-	-	-	-	-	-	-	-	-	-	5.4	5.4
Zimbabwe	-	-	-	-	-	-	-	-	-	-	-	-	-	-	17.8	17.8
Others	0.1	0.2	0.1	0.1	-	-	-	0.3	-	-	-	-	-	-	-	0.8

North America	-	-	6.6	-	-	-	-	-	-	-	-	-	-	-	-	2,386.0
Canada	-	-	-	-	-	-	-	20.2	-	-	-	-	-	-	-	20.2
USA	-	-	6.6	-	-	726.4	-	-	261.9	629.0	56.4	-	-	-	685.5	2,365.8
Central America	0.1	-	0.1	-	-	13.2	-	4.0	28.8	-	-	-	-	-	-	46.2
Costa Rica	-	-	-	-	-	-	-	-	28.8	-	-	-	-	-	-	28.8
Cuba	-	-	-	-	-	13.2	-	-	-	-	-	-	-	-	-	13.2
Guatemala	-	-	-	-	-	-	-	0.1	-	-	-	-	-	-	-	0.1
Jamaica	-	-	-	-	-	-	-	0.1	-	-	-	-	-	-	-	0.1
Mexico	-	-	-	-	-	-	-	3.5	-	-	-	-	-	-	-	3.5
N. Antilles	-	-	0.1	-	-	-	-	0.1	-	-	-	-	-	-	-	0.2
Panama	-	-	-	-	-	-	-	0.2	-	-	-	-	-	-	-	0.2
Others	0.1	-	-	-	-	-	-	-	-	-	-	-	-	-	-	0.1
South America	-	-	-	-	-	14.7	-	2.2	-	13.6	24.0	-	-	-	-	54.5
Argentina	-	-	-	-	-	-	-	2.2	-	-	-	-	-	-	-	2.2
Brazil	-	-	-	-	-	14.7	-	-	-	13.6	-	-	-	-	-	28.3
Chile	-	-	-	-	-	-	-	-	-	-	12.8	-	-	-	-	12.8
Colombia	-	-	-	-	-	-	-	-	-	-	11.2	-	-	-	-	11.2
Asia	12.7	-	7.4	28.6	36.4	60.7	-	316.1	-	30.1	10.4	220.9	145.4	23.0	59.2	950.9
India	12.7	-	-	28.5	-	-	-	11.2	-	10.0	-	146.4	88.1	23.0	21.0	328.2
Israel	-	-	-	-	26.7	-	-	-	-	-	-	-	-	-	-	26.7
Jordan	-	-	7.3	-	9.7	31.9	-	17.3	-	-	10.4	32.8	-	-	-	109.4
Korea, South	-	-	-	-	-	28.8	-	224.9	-	11.5	-	11.2	18.7	-	35.6	330.7
Malaysia	-	-	-	-	-	-	-	-	-	-	-	-	0.6	-	-	0.6
Philippines	-	-	-	-	-	-	-	38.3	-	-	-	14.9	7.9	-	-	61.1
Taiwan	-	-	-	-	-	-	-	24.2	-	8.6	-	15.6	29.8	-	2.6	80.8
Others	-	-	0.1	0.1	-	-	-	0. 2	-	-	-	-	0.3	-	-	0.7
Oceania	-	-	-	-	-	-	-	-	-	-	-	-	-	-	-	-
WORLD	204.8	291.9	538.7	195.5	326.6	2,351.2	220.3	397.4	406.1	965.8	166.1	230.2	145.4	134.0	869.7	7,443.4

a/ Source: British Sulphur Corporation.

ANNEX 11. WORLD UREA TRADE 1984 - TONS OF PRODUCT a/

EXPORTERS	Germany F.R.	Italy	Netherlands	Bulgaria	Czechoslovakia	Germany D.R.	Hungary	Romania	U.S.S.R.	Libya	Indonesia	Kuwait	Qatar	Saudi Arabia	Canada	U.S.A.	Venezuela	Others	Total
IMPORTERS																			
Austria	1,856	225	0	19,634	19,000	0	22,500	6,600	190	0	0	0	0	0	0	0	0	0	70,005
Belgium	7,732	0	2,016	0	0	0	0	0	102	0	0	0	0	0	0	0	0	4,303	14,153
Denmark	22,067	0	134	0	1,560	0	970	0	0	0	0	0	0	0	0	0	0	10,318	35,049
Eire	2	11,040	13,961	0	0	0	0	0	0	0	0	0	0	0	0	0	0	1,389	26,392
France	84,222	9,376	119,731	4,943	0	0	0	0	0	0	0	0	0	0	0	91	0	91,895	310,258
Germany F.R.	0	72	23,306	0	20,321	55,000	11,680	0	0	0	0	0	0	0	0	70	0	40,218	150,667
Italy	8,744	0	0	0	0	0	1,772	18,523	5,216	0	0	0	0	0	18	5,826	0	76,552	116,651
Turkey	0	6,150	0	4,914	0	0	0	69,072	17,195	100,000	0	0	0	0	0	0	0	8,239	205,570
U.K.	9,214	0	31,442	0	1,000	32,394	5,078	0	15,377	0	0	0	0	0	33	44	0	96,716	191,298
Others	36,038	7,379	42,488	1,692	940	7,827	2,615	0	13,451	0	0	0	0	0	46	26	0	99,320	211,822
Total W. Europe	169,875	34,242	233,078	31,183	42,821	95,221	44,615	94,195	51,531	100,000	0	0	0	0	97	6,057	0	428,950	1,331,865
Czechoslovakia	0	0	0	0	0	0	0	0	198,800	0	0	0	0	0	0	0	0	0	198,800
Germany D.R.	0	0	0	0	33,000	0	0	0	0	0	0	0	0	0	0	0	0	15,000	48,000
Hungary	0	0	0	0	0	0	0	0	397,350	0	0	0	0	0	0	0	0	0	397,350
Poland	0	0	0	0	0	0	0	26,000	0	0	0	0	0	0	0	0	0	10,000	36,000
Yugoslavia	3	0	0	3,001	0	0	41,311	0	53,400	0	0	0	0	0	0	0	0	0	97,715
U.S.S.R.	0	0	0	0	0	0	0	0	0	0	0	0	0	0	0	0	0	21,400	21,400
Others	0	0	0	0	0	0	0	0	0	0	0	0	0	0	0	0	0	0	0
Total E. Europe	3	0	0	3,001	33,000	0	41,311	26,000	649,550	0	0	0	0	0	0	0	0	46,400	799,265
Morocco	0	0	0	6,500	5,000	0	0	40,000	29,050	0	0	0	0	0	0	0	0	9,865	90,415
Nigeria	67	0	0	0	0	0	0	100,968	0	0	0	0	0	0	2,544	0	0	4,021	107,600
Sudan	15,250	0	6,721	0	0	0	0	0	0	0	0	0	0	42,382	0	0	0	19,886	84,239
Swaziland	0	0	0	0	0	0	0	10,000	0	0	0	0	0	0	0	0	0	0	10,000
Zambia	0	0	5,000	0	0	0	0	0	0	0	0	0	0	0	0	0	0	18,600	23,600
Zimbabwe	0	0	0	0	0	22,000	0	0	0	0	0	0	0	0	0	0	0	5,770	27,770
Others	12,521	0	35,348	0	0	0	0	24,550	2,250	0	0	3,800	0	1,000	3,015	18,844	0	67,430	168,758
Total Africa	27,838	0	47,069	6,500	5,000	22,000	0	175,518	31,300	0	0	3,800	0	43,382	5,559	18,844	0	125,572	512,382

(Table continues on following page)

WORLD UREA TRADE 1984 - TONS OF PRODUCT a/ (Continued)

EXPORTERS / IMPORTERS	Germany F.R.	Italy	Netherlands	Bulgaria	Czechoslovakia	Germany D.R.	Hungary	Romania	U.S.S.R.	Libya	Indonesia	Kuwait	Qatar	Saudi Arabia	Canada	U.S.A.	Venezuela	Others	Total
Bangladesh	0	0	13,414	0	0	0	0	0	0	0	0	0	0	0	0	83,891	0	51,700	149,005
Burma	14,568	0	0	11,994	0	0	0	0	0	0	10,000	0	0	0	0	0	0	49,140	85,702
China	89,010	58,455	41,000	100,789	61,000	200,000	247,028	454,606	913,117	332,000	0	204,200	195,781	164,073	128,187	296,226	0	903,528	4,389,000
India	52,360	177,328	481,275	183,415	32,000	35,000	0	188,000	331,200	212,000	0	238,400	296,138	177,522	129,290	187,693	0	657,993	3,379,614
Iran	1	0	0	102,548	0	0	0	90,000	142,460	0	0	52,500	57,141	42,500	0	0	0	152,323	639,473
Iraq	0	0	22	0	0	0	0	0	0	0	0	65,300	0	61,476	0	0	0	0	126,798
Malaysia	0	0	13	0	0	0	0	10,000	50,701	0	113,000	0	70,650	0	0	0	0	23,582	267,946
Philippines	0	0	14,716	0	0	0	0	38,600	15,300	0	50,000	0	0	0	0	125,588	0	74,620	318,824
Viet Nam	0	0	0	18,000	0	0	0	33,000	413,600	0	0	0	47,045	29,920	0	0	0	8	541,573
Others	1,962	0	18,896	21,641	0	0	0	300	50,100	15,000	26,000	24,200	5,875	190,376	0	113,689	0	192,751	660,790
Total Asia	157,901	235,783	569,336	438,387	93,000	235,000	247,028	814,506	1,916,478	559,000	199,000	584,600	672,630	665,867	257,477	807,087	0	2,105,645	10,558,725
Australia	3	0	0	0	0	0	0	0	32,100	0	0	0	42,000	0	0	8,536	0	12,007	94,646
Others	1,341	0	0	0	0	0	0	0	0	0	0	0	0	0	0	209	0	1,500	3,050
Total Oceania	1,344	0	0	0	0	0	0	0	32,100	0	0	0	42,000	0	0	8,745	0	13,507	97,696
Canada	0	0	63,925	0	0	0	0	0	0	0	0	0	0	0	0	120,763	0	35,812	220,500
U.S.A.	24,495	54,500	111,451	0	0	102,799	0	356,260	383,239	14,000	0	0	0	0	698,483	0	66,638	106,129	1,917,994
Total N. America	24,495	54,500	175,376	0	0	102,799	0	356,260	383,239	14,000	0	0	0	0	698,483	120,763	66,638	141,941	2,138,494
Argentina	0	10,000	0	20,500	0	0	0	34,000	0	0	0	0	0	0	0	54	9,802	1,350	75,706
Chile	395	0	250	0	0	0	0	0	0	0	0	0	0	0	0	75,177	0	1,100	76,922
Colombia	71	0	0	0	12,000	10,000	5,735	27,266	10,000	0	0	0	0	0	47,498	6,588	141,635	19,439	280,232
Costa Rica	46	0	1,094	0	0	3,000	0	0	0	0	0	0	0	0	5,048	8,879	3,756	4,720	26,543
Cuba	1	0	0	32,000	0	0	0	0	180,000	0	0	0	0	0	0	0	30,000	30,750	272,751
Ecuador	2	0	0	2,000	3,500	0	16,472	30,000	0	0	0	0	0	0	0	13,653	6,903	68	72,598
Guatemala	4,986	0	39,607	0	0	0	0	0	0	0	0	0	0	0	0	13,470	4,434	11,691	74,188
Mexico	0	0	0	0	0	0	0	0	0	0	0	0	0	0	0	0	0	0	0
Nicaragua	0	0	533	0	0	0	0	0	32,200	0	0	0	0	0	7,256	3,497	3,743	12,028	59,257
Peru	0	0	9,949	0	0	11,000	0	10,000	0	0	0	0	0	0	0	21,687	20,063	4,441	77,140
Others	1,695	0	41,789	0	0	14,035	0	6,000	0	0	0	0	0	0	2,816	46,783	12,564	44,858	170,540
Total L. America	7,196	10,000	93,222	54,500	15,500	38,035	22,207	107,266	222,200	0	0	0	0	0	62,618	189,788	232,900	130,445	1,185,877
TOTAL WORLD	388,652	334,525	1,118,081	533,571	189,321	493,055	355,161	1,573,745	3,286,398	673,000	199,000	588,400	714,630	709,249	1,024,234	1,151,284	299,538	2,992,460	16,624,304

a/ Source: Fertecon

ANNEX 12. **WORLD CALCIUM AMMONIUM NITRATE TRADE 1984** - TONS OF PRODUCT a/

EXPORTERS	Austria	Belgium	France	Germany F.R.	Netherlands	Spain	Sweden	U.K.	Germany D.R.	Hungary	Romania	Others	Total
IMPORTERS													
Austria	0	0	13,087	37,877	0	0	0	0	0	12,621	0	0	63,585
Belgium	0	0	81,342	241,323	79,732	220	0	4,067	0	0	0	0	406,684
Eire	0	19,102	0	12,578	36,039	30,082	18,084	9,856	0	0	0	15,054	140,795
France	0	145,836	0	0	95,647	0	2,580	0	0	0	0	0	244,063
Germany F.R.	198,532	399,574	42,236	0	887,058	65,253	33,799	17,556	156,000	66,147	74,649	152,008	2,092,812
Netherlands	0	138,761	123,172	128,104	0	10,144	35,619	65,817	0	0	0	7,602	509,219
Turkey	29,250	0	0	0	0	0	0	0	0	41,446	235,857	144,011	450,564
U.K.	0	34,756	0	17,825	66,253	17,543	33,072	0	0	0	0	123,520	292,969
Others	34,009	3,168	5,851	57,625	16,673	2,603	4,878	0	0	3,872	24,894	51,108	204,681
Total W. Europe	261,791	741,197	265,688	495,332	1,181,402	125,845	128,032	97,296	156,000	124,086	335,400	493,303	4,405,372
Czechoslovakia	12,000	0	0	0	0	0	0	0	0	0	0	0	12,000
Germany D.R.	56,000	0	0	0	0	0	0	0	0	21	0	0	56,021
Others	9,000	0	0	0	0	0	0	0	0	37,743	0	0	46,743
Total E. Europe	77,000	0	0	0	0	0	0	0	0	37,764	0	0	114,764
Total Africa	0	130	199	90,734	50,350	0	0	0	0	0	53,451	80,015	274,879
Total Asia	0	15	168	2,749	3,664	0	0	91	0	0	9,000	0	15,687
Total Oceania	0	957	498	4,456	0	0	0	650	0	0	0	0	6,561
Total N. America	0	0	0	0	23,997	0	0	0	0	0	0	11,000	34,997
Total L. America	0	11	39	1,270	5,284	0	0	0	0	0	0	0	6,604
TOTAL WORLD	338,791	742,310	266,592	594,541	1,264,697	125,845	128,032	98,037	156,000	161,850	397,851	584,318	4,858,864

a/ Source: Fertecon

ANNEX 13. **WORLD AMMONIUM SULPHATE TRADE 1984 - TONS OF PRODUCT a/**

EXPORTERS	Belgium	Germany F.R.	Italy	Netherlands	Bulgaria	Romania	U.S.S.R.	Japan	Korea R.	Canada	U.S.A.	Others	Total
IMPORTERS													
Belgium	0	25,949	0	48,811	0	0	0	0	0	0	0	266	75,026
France	210,652	21,413	11,604	53,811	0	0	0	0	0	0	0	7,342	304,822
Greece	0	12	105,980	9,100	20,967	17,000	31,000	0	0	0	0	5,004	189,063
Spain	23,656	17,148	0	9,051	0	0	0	0	0	0	0	5,600	55,455
Turkey	0	0	141,441	0	99,361	74,000	15,771	0	0	0	42,410	37,017	410,000
U.K.	38,837	12,867	0	28,072	0	0	0	0	0	0	0	3,454	83,230
Others	15,575	41,967	16,416	38,613	0	0	0	0	0	0	176	9,373	122,120
Total W. Europe	288,720	119,356	275,441	187,458	120,328	91,000	46,771	0	0	0	42,586	68,056	1,239,716
Czechoslovakia	0	0	0	0	0	0	144,076	0	0	0	0	0	144,076
Yugoslavia	0	150	0	0	11,796	0	24,924	0	0	0	0	21,726	58,596
Others	0	0	0	0	0	0	0	0	0	0	0	0	0
Total E. Europe	0	150	0	0	11,796	0	169,000	0	0	0	0	21,726	202,672
Cameroon	24,691	6,000	0	6,400	0	0	0	0	0	0	0	0	37,091
Egypt	0	0	0	150	0	0	106,700	0	0	0	0	0	106,850
Libya	0	0	28,615	40	5,300	0	0	0	0	0	0	0	33,955
Morocco	75,680	0	0	0	0	0	0	0	0	0	0	2	75,682
Nigeria	42,004	27,846	0	1,721	0	0	0	0	0	0	0	6	71,577
Tanzania	372	0	0	1	0	0	0	11,052	0	0	0	0	11,425
Others	32,317	2,672	16,456	9,545	0	15,000	0	0	0	0	0	8,068	84,058
Total Africa	175,064	36,518	45,071	17,857	5,300	15,000	106,700	11,052	0	0	0	8,076	420,638

(Table continues on following page)

WORLD AMMONIUM SULPHATE TRADE 1984 - TONS OF PRODUCT a/ (Continued)

EXPORTERS	Belgium	Germany F.R.	Italy	Netherlands	Bulgaria	Romania	U.S.S.R.	Japan	Korea R.	Canada	U.S.A.	Others	Total
IMPORTERS													
China	0	0	0	0	0	11,000	0	79,495	0	0	0	0	90,495
Indonesia	0	1	0	0	0	0	0	10,000	22,298	0	17,406	0	49,705
Iran	0	25,000	0	0	0	0	24,924	0	0	0	0	0	49,924
Malaysia	5,015	5,026	0	0	0	0	0	64,950	7,200	0	7,124	5	89,320
Philippines	0	0	0	0	0	0	0	58,233	32,414	0	51	0	90,698
Sri Lanka	0	5,000	0	0	0	0	0	55,000	0	0	0	5,000	65,000
Thailand	33,668	8,132	0	10,297	0	0	62,600	312,254	32,225	0	0	6,874	466,050
Vietnam	0	0	0	0	0	0	164,900	0	0	0	0	0	164,900
Others	18,732	191	0	5,118	353	0	0	11,187	451	0	30,768	8	66,808
Total Asia	57,415	43,350	0	15,415	353	11,000	252,424	591,119	94,588	0	55,349	11,887	1,132,900
Fiji	0	0	0	0	0	0	0	17,400	30,750	0	0	0	48,150
New Zealand	10,553	180	0	496	0	0	0	20,367	35	2,082	18,413	0	52,126
Others	1,267	1,872	0	36	0	0	0	4,834	0	0	186	0	8,195
Total Oceania	11,820	2,052	0	532	0	0	0	42,601	30,785	2,082	18,599	0	108,471
Canada	0	271	0	624	0	0	0	0	0	0	9,061	0	9,956
U.S.A.	72,455	30,580	0	38,772	0	0	0	0	0	136,736	0	227	278,770
Total N. America	72,455	30,851	0	39,396	0	0	0	0	0	136,736	9,061	227	288,726
Brazil	68,780	55,509	0	118,397	0	0	0	0	0	0	281,350	51,736	575,772
Cuba	0	0	0	0	0	0	389,800	0	0	0	0	0	389,800
Dominican Rep.	3,166	0	0	0	0	0	0	0	0	0	60,550	0	63,716
El Salvador	15,015	10,869	0	5	0	0	0	0	0	0	64,516	1,200	91,605
Guatemala	1,503	3,750	0	20,995	0	0	0	0	0	0	24,654	21,193	72,095
Guyana	0	0	0	0	0	0	0	0	0	0	0	21,000	21,000
Nicaragua	5	8	0	9,952	0	0	0	0	0	0	147	0	10,112
Others	1,936	1,045	0	4,636	500	0	0	0	0	0	54,654	13,021	75,792
Total L. America	90,405	71,181	0	153,985	500	0	389,800	0	0	0	485,871	108,150	1,299,892
TOTAL WORLD	695,879	303,458	320,512	414,643	138,277	117,000	964,695	644,772	125,373	138,818	611,466	218,122	4,693,015

a/ Source: Fertecon

ANNEX 14. WORLD ESTIMATED COMMERCIAL PRODUCTION OF NATURAL GAS AND PROVEN RESERVES
(Billion (Thousand million) cubic meters+/)

	Commercial production a/		Reserves b/		Reserves as % of World c/	Life of reserves Years d/
	1982	1983	1979	1984		
NORTH AMERICA						
Canada	75.85	71.34	1,871	2,613	2.9	37
USA	502.88	450.20	5,716	5,645	6.2	13
	578.73	521.54	7,587	8,258	9.1	16
CARIBBEAN AREA						
Colombia	2.96	2.88	200	117	0.1	46
Mexico	31.64	31.11	1,723	2,180	2.4	70
Trinidad	2.90	3.55	240	430	0.5	+
Venezuela*	17.00	16.25	1,190	1,545	1.7	95
	54.50	53.79	3,353	4,272	4.7	79
OTHER SOUTH AMERICA						
Argentina	11.30	12.55	220	680	0.8	54
Bolivia	2.62	2.60	150	139	0.2	53
Brazil	1.28	1.55	44	82	0.1	53
Chile	1.38	1.34	70	117	0.1	87
Ecuador*	0.07	0.10	50	116	0.1	+
Peru	1.34	1.25	34	25	negl	20
	17.99	19.39	568	1,159	1.3	60
MIDDLE EAST						
Abu Dhabi*	6.97	6.50	780	2,700	3.0	+
Bahrain	3.76	3.95	100	220	0.2	56
Dubai*	1.35	1.29	60	134	0.1	+
Iran*	7.20	8.90	10,700	11,380	12.6	+
Iraq*	0.68	0.65	790	821	0.9	+
Israel	0.16	0.06	1	1	negl	17
Kuwait e/	3.59	4.50	1,120	1,043	1.2	+
Oman	2.70	3.19	100	87	0.1	27
Qatar*	4.92	4.73	312	3,400	3.8	+
Ras-al-Khaiman*	-	-	-	31	negl	na
Saudi Arabia*e/	5.70	5.53	2,105	2,188	2.4	+
Sharjah	-	-	28	285	0.3	na
Syria	0.25	0.25	50	104	0.1	+
Turkey	-	-	1	42	negl	na
	37.28	39.55	16,147	22,436	24.8	+
AFRICA						
Algeria*	26.69	35.59	2,970	3,155	3.5	89
Angola	0.25	0.35	40	53	0.1	+
Cameroun	-	-	-	110	0.1	na
Congo	0.02	0.02	28	70	0.1	+
Cote d'Ivoire	-	-	-	60	0.1	na
Egypt	2.67	3.21	90	200	0.2	62
Gabon	0.15	0.15	68	12	negl	+
Libya*	3.35	4.05	695	555	0.6	+
Morocco	0.08	0.09	1	4	negl	44
Nigeria*	1.42	2.00	1,140	1,370	1.5	+
Rwanda	-	-	-	40	negl	na
Tanzania	-	-	1	118	0.1	na
Tunisia	0.44	0.44	100	85	0.1	+
	35.07	45.90	5,133	5,832	6.5	+
WEST EUROPE						
Austria	1.29	1.23	14	12	negl	10
Denmark	0.03	0.04	97	160	0.2	+
France	6.59	6.66	102	44	negl	7
Germany, F.R.	16.82	17.73	196	195	0.2	11
Ireland	2.05	2.36	30	36	negl	15
Italy	14.59	13.07	200	190	0.2	15
Netherlands	69.73	72.98	1,660	1,927	2.1	26
Norway	24.89	24.42	473	2,039	2.3	83
Spain	-	-	11	25	negl	na
UK	38.28	39.53	705	712	0.8	18
	174.27	178.02	3,488	5,340	5.9	30
FAR EAST						
Afghanistan	2.80	2.85	71	60	0.1	21
Australia	10.97	14.16	848	945	1.0	67
Bangladesh	1.94	2.20	260	340	0.4	+
Brunei	9.21	9.38	220	210	0.2	22
Burma	0.45	0.53	4	210	0.2	+
China	17.65	19.80	700	800	0.9	40
India	2.52	2.99	127	475	0.5	+
Indonesia*	19.08	20.83	1,088	1,000	1.1	48
Japan	2.05	2.12	60	25	negl	12
New Zealand	2.93	3.18	190	152	0.2	48
Taiwan	1.45	1.45	18	23	negl	16
Thailand	1.33	1.61	127	240	0.3	+
	83.70	91.34	4,501	6,390	7.1	70
USSR & EAST EUROPE						
Albania	0.30	0.30	12	8	negl	27
Bulgaria	0.10	0.10	3	5	negl	50
Czechoslovakia	0.63	0.60	15	10	negl	17
Germany P.R.	8.10	7.60	84	60	0.1	8
Hungary	6.63	6.50	100	120	0.1	18
Poland	5.53	5.47	128	115	0.1	21
Romania	39.05	39.60	150	230	0.3	6
USSR	502.00	535.95	27,500	36,000	39.9	67
Yugoslavia	2.19	2.09	62	90	0.1	4.3
	564.53	598.21	28,054	36,638	40.6	61
WORLD TOTAL	1,546.07	1,547.74	68,831	90,325	100.0	58
Of which OPEC countries	98.17	111.07	23,096	29,723	32.9	+

+/ One cubic meter equals 35.3147 cubic feet. For rough equivalent in cubic feet divide bcm production by 10 for billion cubic feet daily and divide bcm reserve by 30 trillion (million million) cubic feet.

*Member of OPEC.

a/ Excluding flared or reinjected; mainly excluding field usage. 1982 production figures are revisions and do not correspond with estimates published in August 1983.

b/ Mainly excluding probable and possible reserves. The five-year comparison with 1979 figures unrevised estimates as published by Petroleum Economist in August 1979.

c/ negl: Negligible (less than 0.05%).

d/ Reserves at 1st January 1984 divided by 1983 rate of commercial production. A plus sign (+) in this column indicates an apparent life greater than 100 years whilst na (not applicable) signifies that commercial production has not yet been established.

e/ Kuwait and Saudi Arabia figures include half shares of Partitioned Zone.

NOTE: Marginal production of associated gas by other countries producing crude oil in 1983 has not been included. Reserves in these and some ten other countries where natural gas finds have been made are principally confined to non-proven.

Source: Cadigas. Petroleum Economist (August 1984).

ANNEX 15. AMMONIA REALIZATION PRICES

A. Variation of IRR % with Realization Prices and Gas Prices

Table 1. Developed Site
Investment US$155 million

Gas Price	Realization Price (US$/ton)											
(US$/MM BTU)	130	150	170	190	210	230	250	270	290	310	330	350
0	11.9	15.1	18.0	20.6	23.1	25.5	27.8	30.0	32.1	34.1	36.0	37.9
1.0	6.0	9.8	13.1	16.1	18.9	21.5	24.0	26.3	28.6	30.7	32.7	34.7
2.0		3.3	7.5	11.1	14.3	17.2	19.9	22.4	24.8	27.1	29.3	31.4
3.0			0.4	5.3	8.9	12.3	15.4	18.2	20.8	23.3	25.6	27.8
4.0					2.3	6.6	10.2	13.5	16.4	19.1	21.7	24.1
5.0							4.1	8.0	11.5	14.6	17.4	20.1

Table 2. Developing Site
Investment US$220 million

Gas Price	Realization Price (US$/Ton)											
(US$/MM BTU)	140	160	180	200	220	240	260	280	300	320	340	360
0	8.1	10.7	13.1	15.4	17.4	19.4	21.3	23.1	24.8	26.5	28.1	29.7
1.0	3.2	6.4	9.1	11.7	14.0	16.1	18.2	20.1	22.0	23.7	25.4	27.1
2.0		1.1	4.5	7.5	10.1	12.5	14.8	16.9	18.9	20.8	22.6	24.3
3.0				2.5	5.7	8.5	11.1	13.4	15.6	17.6	19.6	21.4
4.0						3.8	6.8	9.5	11.9	14.2	16.3	18.3
5.0								5.0	7.9	10.4	12.8	15.0

Table 3. Developing Site, Remote Location
Investment US$270 million

Gas Price	Realization Price US$/ton											
(US$/MM BTU)	190	210	230	250	270	290	310	330	350	370	390	410
0	10.5	12.5	14.3	16.1	17.7	19.3	20.9	22.4	23.7	25.1	26.4	27.7
1.0	7.1	9.2	11.3	13.2	15.0	16.7	18.3	19.8	21.3	22.8	24.2	25.5
2.0	3.0	5.6	7.9	10.0	12.0	13.8	15.6	17.2	18.8	20.4	21.9	23.3
3.0		1.3	4.1	6.5	8.7	10.8	12.7	14.5	16.2	17.8	19.4	20.9
4.0						7.4	9.5	11.5	13.4	15.1	16.8	18.4
5.0							6.0	8.2	10.3	12.2	14.0	15.7

AMMONIA REALIZATION PRICES (Continued)

B. Variation of IRR % with Realization Prices and Utilization Rates

Table 4. Developed Site

Investment US$155 million, Gas Price US$4.0/MMBTU

Plant Utilization	Realization Price (US$/ton)											
(%)	240	250	260	270	280	290	300	310	320	330	340	350
60	1.0	2.6	4.1	5.5	6.9	8.1	9.3	10.5	11.6	12.7	13.8	14.8
70	3.8	5.4	7.0	8.5	9.8	11.2	12.4	13.7	14.8	16.0	17.1	18.1
80	6.3	8.0	9.6	11.1	12.5	13.9	15.2	16.5	17.7	18.9	20.1	21.2
90	8.5	10.2	11.9	13.5	15.0	16.4	17.8	19.1	20.4	21.7	22.9	24.1
100	10.5	12.3	14.0	15.7	17.2	18.7	20.2	21.6	23.0	24.3	25.6	26.8

Table 5. Developing Site

Investment US$220 million, Gas Price US$2.0/MMBTU

Plant Utilization	Realization Price, US$/ton											
(%)	230	240	250	260	270	280	290	300	310	320	330	340
60	4.0	5.0	6.0	7.0	7.9	8.8	9.7	10.1	11.4	12.2	12.9	13.7
70	6.7	7.8	8.8	9.8	10.8	11.8	12.7	13.6	14.4	15.3	16.1	16.9
80	9.1	10.3	11.4	12.4	13.4	14.4	15.4	16.3	17.2	18.1	19.0	19.8
90	11.4	12.6	13.7	14.8	15.9	16.9	17.9	18.9	19.8	20.8	21.7	22.6
100	13.4	14.7	15.8	17.0	18.1	19.2	20.2	21.3	22.3	23.2	24.2	25.1

Table 6. Developing Site, Remote Location

Investment, US$270 million, Gas Price, US$1.0/MMBTU

Plant Utilization	Realization Price (US$/ton)											
(%)	180	200	220	240	260	280	300	320	340	360	380	400
60		1.3	3.2	4.9	6.6	8.1	9.5	10.9	12.2	13.4	14.6	15.8
70	1.7	3.9	5.8	7.6	9.3	10.9	12.4	13.8	15.2	16.5	17.8	19.0
80	3.9	6.1	8.1	10.0	11.8	13.5	15.0	16.6	18.0	19.4	20.7	22.0
90	5.9	8.2	10.3	12.2	14.1	15.8	17.5	19.1	20.6	22.1	23.5	24.9
100	7.7	10.1	12.3	14.3	16.2	18.0	19.8	21.4	23.0	24.6	26.1	27.5

AMMONIA REALIZATION PRICES (Continued)

C. Variation of Realization Price US$/ton with Gas Price and Utilization Rates

Table 7. Developed Site
Investment US$155 million, IRR 15%

Plant Utilization	Gas Price (US$/MMBTU)					
(%)	0	1.0	2.0	3.0	4.0	5.0
60	222	254	287	320	355	385
70	191	223	256	288	321	354
80	168	200	233	266	298	331
90	150	182	215	248	280	313
100	135	168	203	233	266	299

Table 8. Developing Site

Investment US$220 million, IRR 15%

Plant Utilization	Gas Price (US$/MMBTU)					
(%)	0	1.0	2.0	3.0	4.0	5.0
60	292	325	358	390	423	456
70	251	284	317	349	382	415
80	221	253	286	319	352	384
90	197	229	262	295	327	360
100	177	210	243	276	308	341

Table 9. Developing Site, Remote Location

Investment US$270 million, IRR 15%

Plant Utilization	Gas Price (US$/MMBTU)					
(%)	0	1.0	2.0	3.0	4.0	5.0
60	354	387	419	452	485	517
70	304	337	370	402	435	468
80	267	300	332	365	398	430
90	238	270	303	336	369	401
100	214	247	280	313	345	378

AMMONIA REALIZATION PRICES

D. Variation of Realization Price US$/ton with Plant Size and Gas Price

Table 10. Developed Site

IRR 15%, Utilization 90%

Size of Plant & Cost		Gas Price (US$/MM BTU)					
(tpd)	(US$ million)	0	1.0	2.0	3.0	4.0	5.0
750	130	169	201	234	267	299	332
1,000	155	149	182	215	248	280	313
1,250	180	137	169	202	235	268	301
1,500	205	127	160	192	225	258	290

Table 11. Developing Site

IRR 15%, Utilization 90%

Size of Plant & Cost		Gas Price (US$/MM BTU)					
(tpd)	(US$ million)	0	1.0	2.0	3.0	4.0	5.0
750	190	227	260	293	326	358	391
1,000	220	197	229	262	295	327	360
1,250	250	177	210	242	275	308	341
1,500	270	162	194	227	260	292	325

Table 12. Developing Site - Remote Location

IRR 15%, Utilization 90%

Size of Plant & Cost		Gas Price (US$/MM BTU)					
(tpd)	(US$ million)	0	1.0	2.0	3.0	4.0	5.0
750	240	280	313	345	378	411	443
1,000	270	238	271	303	336	369	401
1,250	300	211	244	277	310	342	375
1,500	330	192	224	257	289	323	355

Source: Industry Department
December 1986

ANNEX 16. UREA REALIZATION PRICES

A. Variation of IRR % with Realization Prices and Gas Prices

Table 1. Developed Site
Investment US$230 million

Gas Price	Realization Price (US$/ton)											
(US$/MMBTU)	120	140	160	180	200	220	240	260	280	300	320	340
0	10.1	13.8	17.1	20.1	23.0	25.6	28.1	30.6	32.9	35.1	37.2	39.3
1.0	3.8	8.4	12.3	15.7	18.8	21.7	24.4	27.0	29.5	31.8	34.1	36.2
2.0		1.7	6.6	10.7	14.3	17.5	20.5	23.3	25.9	28.4	30.8	33.0
3.0				4.7	9.0	12.8	16.1	19.2	22.1	24.7	27.3	29.7
4.0					2.7	7.3	11.3	14.7	17.9	20.8	23.6	26.1
5.0						0.5	5.5	9.7	13.3	16.6	19.6	22.4

Table 2. Developing Site
Investment US$320 million

Gas Price	Realization Price (US$/Ton)											
(US$/MMBTU)	150	170	190	210	230	250	270	290	310	330	350	370
0	9.8	12.5	15.0	17.3	19.6	21.6	23.6	25.5	27.3	29.1	30.8	32.5
1.0	5.4	8.6	11.4	14.0	16.4	18.6	20.7	22.7	24.7	26.5	28.3	30.0
2.0		3.9	7.3	10.2	12.9	15.3	17.6	19.8	21.9	23.8	25.7	27.5
3.0			2.4	5.9	9.0	11.8	14.3	16.7	18.9	21.0	33.0	24.9
4.0					4.6	7.8	10.7	13.3	15.7	17.9	20.1	22.1
5.0						3.1	6.5	9.5	12.2	14.7	17.0	19.1

Table 3. Developing Site
Remote Location
Investment US$400 million

Gas Price	Realization Price (US$/ton)											
(US$/MMBTU)	200	220	240	260	280	300	320	340	360	380	400	420
0	12.2	14.2	16.1	17.9	19.7	21.3	22.9	24.5	26.0	27.4	28.9	30.2
1.0	9.0	11.3	13.4	15.3	17.2	18.9	20.6	22.2	23.8	25.3	26.8	28.2
2.0	5.5	8.0	10.3	12.5	14.5	16.4	18.2	19.9	21.5	23.1	24.7	26.1
3.0	1.3	4.3	7.0	9.4	11.6	13.6	15.6	17.4	19.1	20.8	22.4	24.0
4.0			3.1	5.9	8.5	10.7	12.8	14.8	16.6	18.4	20.1	21.7
5.0				1.9	4.8	7.4	9.7	11.9	13.9	15.8	17.6	19.4

B. Variation of IRR % with Realization Prices and Utilization Rates

Table 4. Developed Site

Investment US$230 million, Gas Price US$4.0/MMBTU

Plant Utilization	Realization Price (US$/ton)											
(%)	220	230	240	250	260	270	280	290	300	310	320	330
60	-	1.8	3.4	5.2	6.7	8.1	9.5	10.8	12.0	13.2	14.3	15.4
70	2.7	4.6	6.4	8.1	9.7	11.1	12.6	13.9	15.2	16.5	17.7	18.8
80	5.1	7.1	9.0	10.7	12.3	13.9	15.3	16.8	18.1	19.4	20.7	22.0
90	7.3	9.4	11.3	13.0	14.7	16.3	17.9	19.4	20.8	22.2	23.6	24.9
100	9.3	11.4	13.4	15.2	17.0	18.7	20.3	21.9	23.3	24.8	26.2	27.6

Table 5. Developing Site

Investment US$320 million, Gas Price US$2.0/MMBTU

Plant Utilization	Realization Price (US$/ton)											
(%)	190	200	210	220	230	240	250	260	270	280	290	300
60	-	1.7	3.0	4.2	4.4	6.5	7.5	8.5	9.5	10.5	11.4	12.3
70	2.9	4.3	5.6	6.9	8.1	9.3	10.4	11.4	12.5	13.5	14.4	15.4
80	5.2	6.7	8.0	9.4	10.6	11.8	13.0	14.1	15.2	16.2	17.2	18.2
90	7.3	8.8	10.2	11.6	12.9	14.1	15.3	16.5	17.6	18.7	19.8	20.8
100	9.2	10.8	12.2	13.7	15.1	16.3	17.6	18.8	20.0	21.1	22.2	23.3

Table 6. Developing Site, Remote Location

Investment US$400 million, Gas Price US$1.0/MMBTU

Plant Utilization	Realization Price (US$/ton)											
(%)	190	200	210	220	230	240	250	260	270	280	290	300
60	1.0	2.1	3.1	4.1	5.0	6.0	6.8	7.7	8.5	9.3	10.0	10.8
70	3.5	4.6	5.7	6.7	7.7	8.7	9.6	10.5	11.3	12.1	12.9	13.7
80	5.8	6.9	8.0	9.1	10.1	11.1	12.1	13.0	13.9	14.7	15.6	16.4
90	7.8	9.0	10.2	11.3	12.3	13.4	14.3	15.3	16.3	17.2	18.1	18.9
100	9.7	10.9	12.1	13.3	14.4	15.5	16.5	17.5	18.5	19.4	20.4	21.3

UREA REALIZATION PRICES (Continued)

C. Variation of Realization Price US$/ton with Gas Price and Utilization Rates

Table 7. Developed Site

Investment US$230 million, IRR 15%

Plant Utilization	Gas Price (US$/MMBTU)					
(%)	0	1.0	2.0	3.0	4.0	5.0
60	212	240	268	297	326	354
70	184	213	241	270	298	327
80	164	192	220	249	278	306
90	147	176	204	233	262	290
100	134	163	192	220	249	277

Table 8. Developing Site

Investment US$320 million, IRR 15%

Plant Utilization	Gas Price (US$/MMBTU)					
(%)	0	1.0	2.0	3.0	4.0	5.0
60	276	304	333	361	390	418
70	239	267	296	325	353	382
80	211	240	268	297	326	354
90	190	218	247	276	304	333
100	173	201	230	259	287	316

Table 9. Developing Site, Remote Location

Investment US$400 million, IRR 15%

Plant Utilization	Gas Price (US$/MMBTU)					
(%)	0	1.0	2.0	3.0	4.0	5.0
60	333	361	390	418	447	476
70	288	316	345	374	402	431
80	254	283	311	340	369	397
90	228	251	285	314	342	371
100	207	236	264	293	321	350

Variation of Realization US$/ton Price with Plant Size and Gas Price

Table 10. Developed Site

IRR 15%, Utilization 90%

Size of Plant & Cost		Gas Price (US$/MM BTU)					
(tpd)	(US$ million)	0	1.0	2.0	3.0	4.0	5.0
600	125	228	257	285	314	342	371
1,000	170	181	210	238	267	295	324
1,670	230	147	176	204	233	262	290
2,000	255	135	164	192	221	250	278

Table 11. Developing Site

IRR 15%, Utilization 90%

Size of Plant & Cost		Gas Price (US$/MM BTU)					
(tpd)	(US$ million)	0	1.0	2.0	3.0	4.0	5.0
600	175	291	320	348	377	405	434
1,000	240	233	262	290	319	348	376
1,670	320	190	218	247	276	304	333
2,000	360	174	203	232	260	289	317

Table 12. Developing Site - Remote Location

IRR 15%, Utilization 90%

Size of Plant & Cost		Gas Price (US$/MM BTU)					
(tpd)	(US$ million)	0	1.0	2.0	3.0	4.0	5.0
600	220	348	376	405	434	462	491
1,000	300	280	309	338	367	395	423
1,670	400	228	251	285	314	342	371
2,000	445	207	236	264	293	321	350

Source: Industry Department
December 1986

ANNEX 17. INVESTMENT AND REALIZATION PRICE DATA FOR POTENTIAL UREA PROJECTS

	North America		Western Europe			Oceania	Other Devd.
	USA	Canada	Netherlands	Norway	UK	Australia	Japan
Investment (US$ million)							
(a) New Site	220	250	230	230	230	260	220
(b) Existing Site	176	189	184	184	184	191	176
Gas Price (US$/MMBTU)	2.5	2.0	2.8	3.0	2.8	2.0	3.5
Total Cash Costs (US$/ton)	99	86	109	114	109	86	128
Realization Price (US$/ton)							
10% IRR	170-186	161-183	182-199	187-205	182-199	162-187	199-215
15% IRR	186-213	184-213	205-228	210-233	205-228	185-218	221-242
Production Status/ Major Markets	Domestic	Export NA	Export NA/WE/SEA/FE	Export	Domestic/ Export WE	Domestic/Export SEA/FE	Domestic/ Export FE
Shipping and Terminal Costs (US$/ton)							
North America	-	18	18	18	-	-	-
Western Europe	-	-	13	15	13	-	-
S. E. Asia	-	-	27	27	-	21	-
Far East	-	-	33	33	-	18	14
Landed Cost (10% IRR) (US$/ton)							
North America	-	179-201	200-217	205-223	-	-	-
Western Europe	-	-	195-212	202-220	195-212	-	-
S. E. Asia	-	-	209-226	214-232	-	183-208	-
Far East	-	-	215-232	220-238	-	180-205	213-229
Landed Cost (15% IRR) (US$/ton)							
North America	-	202-231	223-246	228-251	-	-	-
Western Europe	-	-	218-241	225-248	218-241	-	-
S. E. Asia	-	-	232-255	237-260	-	206-239	-
Far East	-	-	238-261	243-266	-	203-236	235-256

INVESTMENT AND REALIZATION PRICE DATA FOR POTENTIAL UREA PROJECTS (Continued)

	Africa						
	Algeria	Angola	Egypt	Libya	Mozambique	Nigeria	Tanzania
Investment (US$ million)							
(a) New Site	360	400	360	360	400	400	400
(b) Existing Site	215	-	215	215	-	-	-
Gas Price (US$/MMBTU)	0.8	0.8	0.8	0.8	0.8	0.8	0.8
Total Cash Costs (US$/ton)	52	52	52	52	52	52	52
Realization Price (US$/ton)							
10% IRR	136-187	202	136-187	136-187	202	202	202
15% IRR	162-226	249	162-226	162-226	249	249	249
Production Status/ Major Markets	Export WE/SEA/FE	Export NA/WE/SEA/FE	Domestic/ Export WE/SEA/FE	Export WE/SEA/FE	Export SEA/FE	Domestic/ Export WE/NA/SEA	Export SEA/FE
Shipping and Terminal Costs (US$/ton)							
North America	-	26	-	-	-	-	-
Western Europe	13	24	15	14	-	22	-
S. E. Asia	27	24	25	26	20	26	20
Far East	33	29	31	32	25	31	25
Landed Cost (10% IRR) (US$/ton)							
North America	-	228	-	-	-	-	-
Western Europe	149-200	226	151-202	150-201	-	224	-
S. E. Asia	163-214	226	161-212	162-213	222	228	222
Far East	169-220	231	167-218	168-219	227	233	227
Landed Cost (15% IRR) (US$/ton)							
North America	-	275	-	-	-	-	-
Western Europe	175-239	273	177-241	176-240	-	271	-
S. E. Asia	189-253	273	187-251	188-252	269	275	269
Far East	195-259	278	193-257	194-258	274	280	274

INVESTMENT AND REALIZATION PRICE DATA FOR POTENTIAL UREA PROJECTS (Continued)

	LATIN AMERICA						
	Argentina	Brazil	Chile	Colombia	Mexico	Trinidad	Venezuela
Investment (US$ million)							
(a) New Site	320	320	360	360	320	320	320
(b) Existing Site	-	208	-	-	208	208	208
Gas Price (US$/MMBTU)	0.8	2.8	0.8	1.0	1.0	1.0	1.0
Total Cash Costs (US$/ton)	52	109	52	58	58	58	58
Realization Price (US$/ton)							
10% IRR	174	191-231	187	193	140-179	140-179	140-170
15% IRR	212	217-271	230	236	165-217	165-217	165-217
Production Status/ Major Markets	Domestic	Domestic	Export NA/FE	Export NA	Domestic/ Export NA/WE	Export NA/WE	Export NA/WE
Shipping and Terminal Costs (US$/ton)							
North America	-	-	24	17	16	17	17
Western Europe	-	-	-	22	22	22	22
S. E. Asia	-	-	-	-	-	-	-
Far East	-	-	30	-	-	-	-
Landed Cost (10% IRR) (US$/ton)							
North America	-	-	211	210	156-196	157-196	157-196
Western Europe	-	-	-	215	162-201	162-201	162-201
S. E. Asia	-	-	-	-	-	-	-
Far East	-	-	217	-	-	-	-
Landed Cost (15% IRR) (US$/ton)							
North America	-	-	254	253	181-233	182-234	182-234
Western Europe	-	-	-	258	187-239	187-239	187-239
S. E. Asia	-	-	-	-			
Far East	-	-	260	-			

INVESTMENT AND REALIZATION PRICE DATA FOR POTENTIAL UREA PROJECTS (Continued)

	Near East						
	Gulf Emirates	Kuwait	Iran	Iraq	Turkey	Syria	Saudi Arabia
Investment (US$ million)							
(a) New Site	360	360	360	360	320	360	360
(b) Existing Site	215	215	215	215	208	215	215
Gas Price (US$/MMBTU)	0.8	0.8	0.8	0.8	2.8	0.8	0.8
Total Cash Costs (US$/ton)	52	52	52	52	109	52	52
Realization Price (US$/ton)							
10% IRR	136-187	136-187	136-187	136-187	191-231	136-187	136-187
15% IRR	162-226	162-226	162-226	162-226	217-274	162-226	162-226
Production Status	Export	Export	Domestic	Export	Domestic	Domestic/ Export	Export
Major Markets	WE/SEA/FE	WE/SEA/FE		WE/SEA/FE		WE/SEA/FE	WE/SEA/FE
Shipping and Terminal Costs (US$/ton)							
North America	-	-	-	-	-	-	-
Western Europe	18	18	-	18	-	18	18
S. E. Asia	18	18	-	18	-	18	18
Far East	24	24	-	24	-	24	24
Landed Cost (10% IRR) (US$/ton)							
North America	-	-	-	-	-	-	-
Western Europe	154-205	154-205	-	154-205	-	154-205	154-205
S. E. Asia	154-205	154-205	-	154-205	-	154-205	154-205
Far East	160-211	160-211	-	160-211	-	160-211	160-211
Landed Cost (15% IRR) (US$/ton)							
North America	-	-	-	-	-	-	-
Western Europe	180-244	180-244	-	180-244	-	180-244	180-244
S. E. Asia	180-244	180-244	-	180-244	-	180-244	180-244
Far East	186-250	186-250	-	186-250	-	186-250	186-250

INVESTMENT AND REALIZATION PRICE DATA FOR POTENTIAL UREA PROJECTS (Continued)

	South East Asia				Far East		
	Bangladesh	Burma	India	Pakistan	Indonesia	Malaysia	Thailand
Investment (US$ million)							
(a) New Site	360	370	360	360	350	350	350
(b) Existing Site	215	218	215	215	214	214	-
Gas Price (US$/MMBTU)	1.0	1.0	2.8	1.0	1.0	1.0	3.0
Total Cash Costs (US$/ton)	58	58	109	58	58	58	114
Realization Price (US$/ton)							
10% IRR	142-193	143-194	194-245	142-193	140-189	140-189	246
15% IRR	168-236	169-240	220-288	168-236	166-231	166-231	288
Production Status/ Major Markets	Domestic/ Export SEA/FE	Domestic/ Export SEA/FE	Domestic	Domestic	Domestic/ Export SEA/FE	Domestic/ Export SEA/FE	Domestic
Shipping and Terminal Costs (US$/ton)							
North America	-	-	-	-	-	-	-
Western Europe	-	-	-	-	-	-	-
S. E. Asia	15	15	-	-	20	20	-
Far East	20	20	-	-	15	15	-
Landed Cost (10% IRR) (US$/ton)							
North America	-	-	-	-	-	-	-
Western Europe	-	-	-	-	-	-	-
S. E. Asia	157-208	158-209	-	-	160-209	160-209	-
Far East	162-213	163-214	-	-	155-204	155-204	-
Landed Cost (15% IRR) (US$/ton)							
North America	-	-	-	-	-	-	-
Western Europe	-	-	-	-	-	-	-
S. E. Asia	183-251	184-255	-	-	186-251	186-251	-
Far East	188-256	189-260	-	-	181-246	181-246	-

INVESTMENT AND REALIZATION PRICE DATA FOR POTENTIAL UREA PROJECTS (Continued)

	Eastern Europe		Socialist Asia
	USSR	Romania	China
Investment US$ (million)			
(a) New Site	360	320	360
(b) Existing Site	215	208	215
Gas Price (US$/MMBTU)	1.0	2.5	1.7
Total Cash Costs (US$/ton)	58	100	80
Realization Price (US$/ton)			
10% IRR	142-193	182-222	162-199
15% IRR	168-236	207-260	188-237
Production Status/ Major Markets	Domestic/ Export NA/WE/SEA/FE	Domestic/ Export NA/ WE/SEA/FE	Domestic
Shipping and Terminal Costs (US$/ton)			
North America	20	20	-
Western Europe	13	13	-
S. E. Asia	27	27	-
Far East	33	33	-
Landed Cost (10% IRR) (US$/ton)			
North America	162-213	202-242	-
Western Europe	155-206	195-235	-
S. E. Asia	169-220	209-249	-
Far East	175-229	215-255	-
Landed Cost (15% IRR) (US$/ton)			
North America	188-256	227-280	-
Western Europe	181-249	220-273	-
S. E. Asia	195-263	234-287	-
Far East	201-269	240-293	-

Source: Industry Department
December 1985

ANNEX 18. INVESTMENT AND REALIZATION PRICE DATA FOR POTENTIAL AMMONIA PROJECTS

	North America		Western Europe	Africa				
	USA	Canada	Netherlands	Libya	Algeria	Mozambique	Nigeria	Tanzania
Investment (US$ million)								
(a) New Site	145	158	155	240	240	270	270	270
(b) Existing Site	116	126	124	147	147	-	-	-
Gas Price (US$/MMBTU)	2.5	2.0	2.8	0.8	0.8	0.8	0.8	0.8
Total Cash Costs (US$/ton)	100	84	109	46	46	46	46	46
Realization Price (US$/ton)								
10% IRR	175-195	165-189	189-209	139-197	139-197	215	215	215
15% IRR	199-222	190-222	215-240	168-244	168-244	268	268	268
Production Status/ Major Markets	Domestic	Export NA	Export	Export	Export	Export SEA/FE	Domestic/Export NA/WE	Export SEA/FE
Shipping and Terminal Cost (US$/ton)								
North America	-	30	30	35	35	-	40	-
Western Europe	-	-	-	20	20	-	40	-
S.E. Asia	-	-	40	35	35	35	-	35
Far East	-	-	-	50	50	40	-	40
Landed Cost (10% IRR) (US$/ton)								
North America	-	195-219	219-239	174-232	174-232	-	255	-
Western Europe	-	-	-	159-217	159-217	-	255	-
S. E. Asia	-	-	229-249	174-232	174-232	250	-	250
Far East	-	-	-	189-247	189-247	255	-	255
Landed Cost (15% IRR) (US$/ton)								
North America	-	220-252	245-270	203-279	203-279	-	308	-
Western Europe	-	-	-	188-264	188-264	-	308	-
S. E. Asia	-	-	255-280	203-279	203-279	303	-	303
Far East	-	-	-	218-294	218-294	308	-	308

INVESTMENT AND REALIZATION PRICE DATA FOR POTENTIAL AMMONIA PROJECTS (Continued)

	Latin America			Near East		Far East	Eastern Europe
	Trinidad	Mexico	Venezuela	Gulf Emirates	Saudi Arabia	Indonesia	USSR
Investment (US$ million)							
(a) New Site	220	220	220	240	240	230	240
(b) Existing Site	144	144	144	147	147	144	147
Gas Price (US$/MMBTU)	1.0	1.0	1.0	0.8	0.8	1.0	1.0
Total Cash Costs (US$/ton	52	52	52	46	46	52	52
Realization Price (US$/ton)							
10% IRR	144-192	144-192	144-192	139-197	139-197	145-197	145-203
15% IRR	172-234	172-234	172-234	168-244	168-244	173-242	175-250
Production Status/ Major Markets	Export NA/WE	Domestic/Export NA/WE	Export NA/WE	Export WE/SEA/FE	Export WE/SEA/FE	Domestic/Export SEA/FE	Domestic/Export NA/WE/SEA/FE
Shipping and Port Costs (US$/ton)							
North America	20	20	20	-	-	-	35
Western Europe	30	30	30	40	40	-	25
S.E. Asia	-	-	-	20	20	25	35
Far East	-	-	-	35	35	15	40
Landed Cost (10% IRR) (US$/ton)							
North America	164-212	164-212	164-212	-	-	-	180-238
Western Europe	174-222	174-222	174-222	179-237	179-237	-	170-228
S. E. Asia	-	-	-	159-217	159-217	170-222	180-238
Far East	-	-	-	174-232	174-232	160-202	185-243
Landed Cost (15% IRR) (US$/ton)							
North America	192-254	192-254	192-254	-	-	-	210-285
Western Europe	202-264	202-264	202-264	208-284	208-284	208-284	200-275
S. E. Asia	-	-	-	188-264	188-264	188-264	210-285
Far East	-	-	-	203-279	203-279	203-279	215-290

Source: Industry Department
December 1986

ANNEX 19. WORLD GRAIN PRODUCTION AND NITROGEN CONSUMPTION

Year	World Population (000)	Fertilizer N. Consumption (000' tons)	Total Grain Production (000'tons)	N Per Capita MT	Grain Production Per Capita (.001 MT)	Grain/Nitrogen
1950	2,504,203	3,940.3	720,000	1.57	0.29	182.73
1951	2,550,718.6	4,330.0	735,000	1.70	0.29	169.75
1952	2,598,098.2	4,751.6	755,000	1.83	0.29	158.89
1953	2,646,357.8	5,214.2	775,000	1.97	0.29	148.63
1954	2,695,513.9	5,721.8	770,000	2.12	0.29	134.57
1955	2,745,547.0	6,278.9	818,000	2.29	0.30	130.28
1956	2,797,218.2	6,890.2	868,000	2.46	0.31	125.98
1957	2,849,861.8	7,561.1	855,000	2.65	0.30	113.08
1958	2,903,496.2	8,297.2	935,000	2.86	0.32	112.69
1959	2,958,140.0	9,105.0	930,000	3.08	0.31	102.14
1960	3,013,816.0	9,991.5	952,000	3.32	0.32	95.28
1961	3,073,379.6	10,964.3	924,000	3.57	0.30	84.27
1962	3,134,120.3	12,031.7	980,912	3.84	0.31	81.53
1963	3,196,061.5	12,334.4	979,165	3.86	0.31	79.38
1964	3,259,226.8	13,876.9	1030,398	4.26	0.32	74.25
1965	3,323,641.0	15,235.0	1,025,941	4.58	0.31	67.34
1966	3,392,676.3	17,278.9	1,090,160	5.09	0.32	63.09
1967	3,463,145.6	22,091.7	1,134,136	6.38	0.33	51.34
1968	3,535,078.6	24,357.4	1,17,1525	6.89	0.33	48.10
1969	3,608,505.7	26,471.9	1,180,565	7.34	0.33	44.60
1970	3,683,456.0	28,362.8	1,204,622	7.70	0.33	42.47
1971	3,758,826.9	31,796.1	1,310,711	8.46	0.35	41.22
1972	3,835,740.0	33,386.5	1,272,859	8.70	0.33	38.12
1973	3,914,226.9	3,6045.8	1,375,143	9.21	0.35	38.15
1974	3,994,319.8	3,8991.4	1,338,457	9.76	0.34	34.33
1975	4,076,045.0	3,8503.9	1,372,346	9.45	0.34	35.64
1976	4,148,154.3	4,3575.8	1,481,585	10.50	0.36	34.00
1977	4,221,539.3	4,5223.5	1,472,089	10.71	0.35	32.55
1978	4,296,222.6	4,9009.3	1,604,617	11.41	0.37	32.74
1979	4,372,227.0	5,3838.8	1,556,108	12.31	0.36	28.90
1980	4,449,568.0	5,7168.5	1,568,055	12.85	0.35	27.43
1981	4,530,888.3	6,0726.8	1,653,454	13.40	0.36	27.23
1982	4,613,694.8	6,0309.7	1,702,940	13.07	0.37	28.24
1983	4,698,014.7	6,1020.7	1,644,421	12.99	0.35	26.95
1984	4,783,875.6	6,6880.0	1,801,684	13.98	0.38	26.94

Source: Industry Department
February 1986

ANNEX 20. COUNTRY AND REGIONAL NITROGEN PROFILES

1.01 This section briefly describes the main features of nitrogen demand, supply and trade for those countries which are involved to any significant extent in the world nitrogen business. All pertinent country data are given in Annexes 1-6 and not necessarily included in the country profiles. Annexes 1 and 2 contain country capacity and supply capability tables. On the demand side, both fertilizer and industrial nitrogen historic consumption and future demand are given on a country or regional basis in Annexes 4, 5 and 6.

1.02 Generally it has been the intention to conform to the FAO data on demand but in some cases where FAO information is based on apparent demand, (i.e., production plus imports less exports) and more precise data are available on actual demand from other reliable sources, the latter information has also been stated. A word of caution is also necessary to differentiate between the forms of N that are used in making nitrogen balances. Normally, world and regional nitrogen balances are based on ammonia, because almost all nitrogen products are produced from ammonia. On the other hand ammonia, although often used directly as a fertilizer, is not always included in the FAO statistics on production, consumption and trade, where usually only statistics on finished nitrogen products are given. Where a country is involved in ammonia production and trade to a significant extent this information has also been included in the country profiles as well as in Section V which deals with trade. Because of stock changes from year to year the tables on fertilizer production, consumption and trade may not always balance. Fertilizer information from FAO and many other sources is usually presented on a fertilizer split year basis, (i.e., July 1-June 30). For countries that report their fertilizer statistics on a calendar year basis data are shown under the fertilizer year, the first half of which corresponds to the calendar year; i.e., 1984 data are under 1984/85.

A. Developed Market Economies

1. North America

1.03 North America, next to Eastern Europe and Centrally Planned Asia, is the third largest regional producer and consumer of nitrogen products. The region is dominated by the USA which in 1984/85 accounted for about 80% of the region's production and 90% of consumption.

Canada

1.04 Nitrogen fertilizer consumption in Canada grew rapidly throughout the 1970's averaging an increase of almost 10% py. This was due mainly to increased fertilizer demand in the Prairy provinces to increase grain production for the export market. Consumption stagnated in 1981/82 and to a lesser extent in 1982/83 but recovered to show about 15% increase in nitrogen consumption in 1983/84 and a further 12% in 1984/85. The low prices and strong competition in the international grain markets are likely however to constrain the increase in demand rate in the future to about 3% py.

1.05 Canada has a strong and growing nitrogen production industry supported by large reserves of natural gas and is already a large exporter of nitrogen fertilizers. In 1984 Canada exported 1.02 million tons of urea and about 0.66 million tons of ammonia, mainly to the USA. In 1985 there were nine companies in Canada operating installed ammonia capacity equivalent to about 3.0 million tons py N. Several new plants are being constructed which will increase the ammonia capacity in Canada by about 1 million tons by 1990, much of which will be exported to the USA. Amongst the new plants, and which came on-stream at the end of 1985, is that of C-I-L which uses the new Imperial Chemical Industries (ICI) AMV process and which is expected to be one of the most energy efficient ammonia plants in the world.

Ammonia Exports from Canada ('000 tons N)

1980	1981	1982	1983	1984
354	383	426	553	684

Canada

Nitrogen Fertilizer Production, Consumption and Trade ('000 tons N)

	1979/80	1980/81	1981/82	1982/83	1983/84	1984/85
Production	1,709	1,755	1,825	1,897	2,371	2,543
Consumption	831	914	966	1,002	1,157	1,255
Exports	924	994	965	994	1,217	1,511
Imports	136	99	126	145	190	190

1.06 Canada has very large reserves of natural gas, mainly in Alberta and British Columbia, but also to a lesser extent in Saskatchewan and Eastern Canada. Much of the new development based on natural gas, in both Alberta and British Columbia, is directed to the USA and other overseas markets. Gas prices east of Alberta are currently determined by the Canada/Alberta agreement which relates the wholesale price of gas at the Toronto city gate to 65% of the wholesale price of crude oil at the Toronto refinery gate on a heat equivalent basis. Other aspects of the agreement relate gas east of Alberta to the Alberta border price of natural gas plus the cost of transport and distribution and taxes. The cost of gas in Alberta itself is determined largely by the market and is normally up to 30% lower than the "Border Price."

United States of America

1.07 Until about 1981 the USA was the world's largest producer of ammonia and nitrogen fertilizers. The availability of relatively cheap natural gas, low investment costs for production units and the introduction of high yield varieties of corn together with the export boom of 1970s had stimulated the growth of the US nitrogen industry. Between 1960 and 1980 ammonia capacity increased from 4 million tons to almost 16 million tons. After 1980 several factors have adversely affected the US nitrogen industry. A world recession, which affected demand for US agricultural products, high interest rates and a highly valued US dollar caused a sharp fall in US nitrogen fertilizer demand. At the same time increasing variable costs due to rising gas prices made it difficult to compete with imported ammonia. About 5 million tons of ammonia capacity out of a total of 16 million tons was idled for various periods between 1981-1985 and about 3 million tons will remain permanently closed. Within this period the USA changed from being a net exporter of nitrogen to a net importer of nitrogen.

1.08 In 1983/84 after two years of uncertainty in the industry caused by the PIK (Payment in Kind) Program and the drought, nitrogen fertilizer demand in the US increased by 20% from its low level of 8.28 million tons in 1982/83 to 10.11 million tons in 1983/84.

1.09 In 1985, with record crops and a relatively low demand for agricultural products, the US farming industry was facing even a worse crisis than that which produced the PIK program in 1983. In December 1985 the Food Security Act of 1985 (also known as the "New Farm Bill") was signed. It replaced earlier legislation including the PIK program. There seems little doubt that the new bill which is designed to protect farm income and improve market competitiveness will result in considerable acreage reduction. This together with the large existing stocks of cereal and coarse grain stocks is likely to influence fertilizer consumption adversely. Nitrogen demand is expected to decline in the USA by about 10% during 1986 and 1987.

1.10 There are more than forty producers of ammonia in the USA, with a capacity of about 13.9 million tons in 1984/85 as indicated on the following page.

Company	Ammonia Capacity (million tons N)
Agrico Chemicals	1.30
Arcadian Corporation	0.38
C. F. Industries	1.29
Farmland Industries	1.60
IMC Chemical Company	0.57
Monsanto Company	0.47
American Cyanamid	0.43
Union Chemical Company	0.97
Others	6.87
Total	13.88

1.11 There are no significant planned expansions of capacity in the USA through the remainder of the decade. The experiences of the early 1980s with very low import prices, which caused many producers to idle their plants, will deter many producers from future investments. Other producers believe that only those projects with low cash costs can weather the price fluctuations of the international ammonia market and are looking at overseas projects with low gas prices which would be competitive in the US market. Others are looking to Canada as a future source of production. However, the recent turn-down in oil prices which has also resulted in reduced gas prices means that the relative competitive position of domestic production compared with imports will improve so no further major closures of ammonia capacity in the USA are likely through 1990.

1.12 The USA itself has been a net importer of anhydrous ammonia for several years but much of this imported ammonia was processed and exported as diammonium phosphate. In 1982, the USA experienced an overall nitrogen deficit when consumption of nitrogen exceeded domestic production.

Ammonia Imports & Exports USA ('000 tons N)

	1980	1981	1982	1983	1984
Imports	1,739	1,570	1,184	1,964	2,366
Exports	618	461	553	272	397

1.13 Imported ammonia comes mainly from Canada, Mexico, the USSR and Trinidad. Until recently the USA was expected gradually to increase its N deficit throughout the second half of the 1980s. However, with the reduced crop acreage and fertilizer consumption, this now seems unlikely, and domestic supply capability should be more than adequate to meet demand, although a balanced nitrogen situation should be reached again by the end of the decade as demand picks-up again.

USA

Nitrogen Fertilizer Production Consumption and Trade ('000 tons N)

	1979/80	1980/81	1981/82	1982/83	1983/84	1984/85
Production	11,180	11,825	10,523	9,000	9,401	11,121
Consumption	10,348	10,817	9,904	8,280	10,063	10,436
Exports	2,396	2,800	2,267	1,886	1,889	2,932
Imports	2,370	2,227	2,296	2,427	3,574	3,312

1.14 The USA has large reserves of natural gas and until 1983 was the largest world producer of gas. However in 1983, the US gas production dropped to its lowest level for 18 years and at the same time production in the USSR rose by about 7% to make it the world's largest producer and consumer of natural gas. Throughout the 1970s, proven reserves of natural gas in the USA had fallen steadily but in 1980 following the Natural Gas Policy Act (NGPA) of 1978 and the Middle East energy crisis of 1979, increased exploration in the USA resulted in a small increase in reserves. Usage of natural gas in the USA is high - of the order of 450-500 billion cubic meters (BCM) per year and most forecasters feel it will be difficult to find new reserves of gas to keep up with this usage.

1.15 The gas situation in the USA is complex and there are widely differing views on the impact of the NGPA and also on the forecasts for the long-term pricing of gas. Some analysts believe that gas supplies will exceed premium markets until the year 2000, particularly with increasing gas available at favorable terms from Canada and Mexico. Others see gas prices rising more rapidly as the economy, and hence demand for gas, improves. Natural gas prices for the large ammonia market are critical and in 1985 contracts covered a wide range of gas prices from less than $0.25 up to more than $5.0, but with most gas prices being in the range $2.5-3.5/MMBTU range as indicated below. Table 16 (Section VI) shows how gas prices in the US fertilizer industry have changed over the last decade.

US Gas Costs US$/MMBTU - The Fertilizer Institute (TFI) Survey of 27 Companies 1984 a/

Under $0.50	$0.50-1.49	$1.5-2.49	$2.5-3.49	>$3.50
12%	12%	-	56%	20%

a/ Representing about 90% of US industry.

1.16 Industrial gas prices after deregulation in 1985 were expected to move gradually to parity with industrial No. 6 fuel oil. The large fall in oil prices during the first half of 1986 obviously had a major impact on the projections of gas prices which are now in the order of $2-2.5/MMBTU by 1990.

2. Western Europe

1.17 Traditionally, West Europe has been one of the largest producing and consuming regions of nitrogen fertilizer. Up to 1960, as elsewhere, the European nitrogen fertilizer industry was characterized by a large number of small scale production units using an assortment of fuels. In the mid-1960s natural gas became available in the Netherlands in large quantities for industrial use and this was followed by the discovery and subsequent exploitation of natural gas reserves in other parts of Western Europe such as the North and Irish Seas. About 40 new large ammonia plants were built in Europe between 1965-1975. During this period nitrogen fertilizer consumption also increased rapidly up to 1980, after which it declined due to the world recession. The market for nitrogen fertilizers is now expected to grow at less than 1% py over the next decade and the region will remain reasonably well balanced with regard to supply and demand. Seventeen of the twenty countries classified as Western Europe are producers of ammonia but the major producing and consuming countries are France, Netherlands, Italy, Germany and the UK. The future of the Western European nitrogen fertilizer industry will depend to a large extent on the Agricultural Policy of the European Economic Community (EEC) particularly with regard to grass (dairy products) and cereal production where nitrogen fertilizer is a major input.

Germany F.R.

1.18 Domestic nitrogen fertilizer production in West Germany has declined significantly in the last ten years and following several plant closures recently, capacity has dropped by about 25% since 1979/80. The main reasons for this is a mature nitrogen market, the depressed economy of recent years and the relatively high energy prices in Germany. Production and marketing are dominated by BASF and to a lesser extent Ruhr-Stickstoff (now part of Norsk Hydro). Although BASF owns or controls more than half of German ammonia capacity it also has interests in Belgium, France and the Netherlands.

1.19 There is intense competition in the German market from other Western European Countries and also from new plants in Eastern Europe and it seems likely that in the face of a low growth rate of demand, ammonia and nitrogen fertilizer plant capacity may decline further.

Germany, F.R.

Nitrogen Fertilizer Production, Consumption and Trade ('000 tons N)

	1979/80	1980/81	1981/82	1982/83	1983/84	1984/85
Production	1,476	1,436	1,108	985	1,111	1,195
Consumption	1,477	1,551	1,323	1,464	1,378	1,452
Exports	454	424	362	454	508	521
Imports	561	529	540	816	684	673

1.20 Germany has limited natural gas reserves (though it imports gas from the Netherlands and USSR) and about 60% of ammonia capacity is based on gas. The remainder (Brunsbuttel and Gelsenkirchen) uses heavy fuel oil.

Italy

1.21 Italy is a large producer of nitrogen fertilizers, mainly through the two major companies ENICHEM Agricultura SpA - part of the ENI Group - and Fertimont, a subsidiary of Montedison. ENICHEM Agricultura has a capacity of about 0.7 million tpy N as ammonia with facilities to convert this into urea, ammonium nitrate and complex fertilizers in three major sites at Gela, Manfredonia-Apulia and Ravenna. Montedison operates about 1 million tpy nitrogen capacity and accounts for about 48% of the total production of nitrogen fertilizers. Both production and consumption had fallen steadily in Italy between 1979 and 1984 but they recovered somewhat in 1984/85. As demand in Italy is well below that in many other European countries it is expected to increase significantly in the future.

1.22 Italy has also been a major exporter of nitrogen fertilizers but this business is likely to decline due to intense competition from production sources with more favorable gas prices.

Italy

Nitrogen Fertilizer Production, Consumption and Trade ('000 tons N)

	1979/80	1980/81	1981/82	1982/83	1983/84	1984/85
Production	1,523	1,318	1,134	1,116	1,081	1,338
Consumption	1,107	1,006	988	968	996	1,017
Exports	453	543	282	217	334	459
Imports	236	176	164	194	230	194

1.23 The Italian nitrogen industry is based mainly on natural gas. About one-third of the gas used in Italy comes from domestic resources, the remainder comes from Algeria through the Trans-Mediterranean pipeline or from the USSR or the Netherlands.

United Kingdom

1.24 Although the UK has a well-established nitrogen fertilizer industry based on North Sea gas, it has also become increasingly dependent on imports of nitrogen fertilizers to meet its needs. The nitrogen industry in the UK is dominated by ICI with about 85% of domestic ammonia capacity. ICI claims to be the largest private sector ammonia producer in the world. Most of the remaining production comes from a subsidiary company of the Dutch company UKF based in the UK. Recently another major world nitrogen producer entered the market when Norsk Hydro took over Fisons Fertilizers Limited. The consumption of nitrogen fertilizers has increased steadily over the last few years after a setback in 1981. With improved types of wheat and high fertilizer usage, yields of cereal in the UK are now amongst the highest in the world and the reduced cost of production has allowed the UK to become a competitive exporter of cereals. Although at a reduced rate, compared to the past, nitrogen fertilizer growth in the UK is expected to continue at a higher rate than most other European countries.

UK

Nitrogen Fertilizer Production and Trade ('000 tons N)

	1979/80	1980/81	1981/82	1982/83	1983/84	1984/85
Production	1,316	1,167	1,270	1,399	1,372	1,360
Consumption	1,314	1,240	1,386	1,560	1,601	1,580
Exports	92	121	122	110	153	139
Imports	189	162	222	289	438	360

1.25 All the UK's nitrogen industry is based on natural gas supplied from the North Sea through contracts with British Gas.

France

1.26 The French fertilizer industry has undergone major restructuring during the last few years and is now dominated by two organizations, Compagnie Francaise de l'Azote (COFAZ), which until recently was a subsidiary of Parisbas, and CdF Chemie AZF, a CdF Chemie subsidiary. Rhone-Poulenc, which was France's largest fertilizer group, has given up its fertilizer operations. Its subsidiaries Generale des Engrais (GESA) and Societe des Participations Gardinier (SOPAG) have been transferred to CdF Chemie AZF while COFAZ took over SOPAG. These reorganizations have been made to strengthen the French Industry against imports and improve the profitability of the industry. Norsk Hydro has recently taken over COFAZ and has announced its intention to build a large ammonia plant in France probably at Montoir.

France

Nitrogen Fertilizer Production and Trade ('000 tons N)

	1979/80	1980/81	1981/82	1982/83	1983/84	1984/85
Production	1,780	1,640	1,553	1,530	1,600	1,695
Consumption	2,135	2,146	2,193	2,196	2,320	2,337
Exports	304	250	187	217	250	293
Imports	697	756	827	883	900	951

1.27 Many French ammonia plants ran until recently on naphtha but have now been retrofitted to gas. France has limited reserves of gas and will rely more and more on imports of gas from the USSR and Algeria.

Netherlands

1.28 Based on large reserves of natural gas, the Dutch nitrogen industry has developed to become the major nitrogen producer and exporter in Western Europe. The major producer is Unie Van Kunstmestfabriekenbv (UKF) a/ which was formed by the merger of three major Dutch fertilizer producers in 1972. However in recent years overseas interests have taken a stake in the nitrogen industry in the Netherlands, when Norsk Hydro of Norway took over Nederlandse Stikstof Maatschappij (NSM) and most recently when Kemira Oy of Finland acquired the ammonia plant of Esso Chemie. UKF also has manufacturing facilities in the UK, Belgium, France, Ireland and in the USA. In the face of retractions elsewhere, the Netherlands has an expansionary policy towards new nitrogen production and has recently commissioned two new large nitrogen plants. Until recently, producers in the Netherlands benefitted from a dual gas price system. For product destined for the European market, the gas price was based on the tariff system for large industrial consumers, whereas for product destined for exports outside Europe the gas price was based on levels to be competitive in such markets. Nitrogen consumption in the Netherlands is close to optimum application rates and little growth is foreseen. Increased production will therefore go mainly to exports both within and outside Western Europe. Norsk Hydro has announced its intention to build another large scale plant at N.S.M. Sluiskil.

---

a/ In September 1986 the name was changed to DSM Agro bv.

Netherlands

Nitrogen Fertilizer Production, Consumption and Trade ('000 tons N)

	1979/80	1980/81	1981/82	1982/83	1983/84	1984/85
Production	1,612	1,624	1,463	1,504	1,700	1,739
Consumption	486	483	477	457	478	485
Exports	1,252	1,195	1,092	1,262	1,417	1,493
Imports	104	98	131	166	173	151

1.29 The Netherlands have significant reserves of natural gas and these are being exploited at a high rate, making the Netherlands the third largest producer of natural gas in the world.

Other Western European Countries

1.30 Other major producers of ammonia in Western Europe include Spain, Norway, Belgium, Yugoslavia, Austria and Ireland.

Spain: With capacity and consumption of nitrogen both just over 800,000 tpy of N, the supply/demand situation is fairly well balanced. The Spanish fertilizer sector including ammonia capacity is currently being restructured to meet the anticipated needs of 1990 when demand should reach about 1.3 million tons of N as ammonia, of which about half would be imported. A number of the high cost plants at Huelva, La Coruna, Las Palmas and Malaga will be closed down and the remaining plants will be converted to natural gas which will be imported from Algeria.

Norway: Although demand for nitrogen fertilizers in Norway is relatively small at about 115,000 tpy, it is the base of one of the world's largest nitrogen producers, Norsk Hydro. This company has a nitrogen capacity of about 650,000 tpy from four different sites in Norway producing urea, ammonia nitrate and calcium ammonium nitrate, most of which is exported, mainly to other

Scandinavian countries. The Norwegian nitrogen fertilizer industry was originally based on the electrolysis of water using relatively cheap hydro-electric power and about 30% of the industry still uses this method. About 10% of production is based on coke oven gas and nearly 60% on natural gas. Norway has large reserves of natural gas in the North Sea and has talked about building a large additional capacity at the end of the 1980s.

Belgium: With an annual consumption of about 200,000 tpy and a capacity of 570,000 tpy, Belgium is a major exporter of nitrogen fertilizers. The Belgian nitrogen is dominated by two producers - Societe Carbochimique of Belgium and BASF of West Germany. It seems unlikely that there will be any increase in capacity in the near future in view of declining export prospects. Following the closure of the small plant at Willebroek and Marley which were based on heavy fuel oil, the other plants in Belgium at Tertre and Ostend are based on natural gas which is imported from Holland.

Yugoslavia: Although coming within the FAO classification as Western Europe this country is more closely linked with the East European industry. Nitrogen fertilizer demand has increased steadily over the last two decades and is expected to increase at 3% py over the next five years. Until the early 1980s Yugoslavia was a net importer of nitrogen but with a large plant that came on-stream at Kutina in 1983/84 and another one due to come on-stream at Pancevo in 1985/86 the country will have an export potential of about 300,000 tons of nitrogen by the end of the decade. About 10% of the capacity is based on coke oven gas, mainly the old small plants, but all the new capacity is based on natural gas from domestic reserves.

Austria: Austria consumes about 160,000 tpy N and that is expected to increase demand at about 2.5% py throughout the decade to about 180,000 tpy N. This is well below half of Austrian capacity for producing nitrogen fertilizers and the country exports about 300,000 tpy of nitrogen products. The Austrian nitrogen fertilizer industry originated as part of the iron and steel complex at Linz, using coke oven gas. The company now operates two ammonia plants based on imported natural gas which comes from Western Europe and the USSR. With relatively high gas cost there seems little likelihood of increased ammonia production in Austria and it also seems likely that exports will decline in the face of severe competition from Eastern Europe.

Ireland: The nitrogen fertilizer market has grown steadily in Ireland and consumption was about 300,000 tpy in 1983/84 and is expected to increase to about 400,000 tpy by the end of the decade. Until about 1979, Ireland depended on importing a good part of its nitrogen needs but following the commissioning of a large new plant at Arklow, the country is now a net exporter of nitrogen. Nitrigin Eireann Teoranta (NET) is the sole producer and operates the nitrogen complex at Marina Point which has a capacity of 495,000 tpy ammonia and 320,000 tpy urea as products. Natural gas for the plant is purchased from the Irish Gas Board and comes from the Marathon field in the Irish Sea. Statistics on other countries in Western Europe are given in Annexes 1-6.

3. Oceania

Australia

1.31 Consumption of nitrogen fertilizers in Australia has always lagged well behind phosphate fertilizers, due mainly to the large input of nitrogen through clover and other legume-based pastures. This is changing and there has been a marked increase in cereal cropping with less reliance on nitrogen produced from clover and more from chemical nitrogen fertilizers. There has been little increase in Australian nitrogen capacity over the last ten years and with increasing demand, the country has been a net importer since 1980. The industry has a nominal capacity of about 450,000 tpy N based mainly on natural gas. About 25% of the production from these plants goes to industrial use.

Australia

Nitrogen Fertilizer Production, Consumption and Trade ('000 tons N)

	1979/80	1980/81	1981/82	1982/83	1983/84	1984/85
Production	210	214	206	197	173	214
Consumption	245	248	250	273	269	320
Exports	5	11	8	5	1	1
Imports	28	45	52	79	64	116

1.32 Ammonia plants in Australia use both naphtha and natural gas. New capacity however will be based on natural gas from off-shore gas in Western Australia or in Queensland.

New Zealand

1.33 The pattern of fertilizer consumption in New Zealand has been similar to that in Australia with relatively small quantities of nitrogen used compared with phosphates. A small ammonia/urea complex with 76,000 tpy N as ammonia and 73,000 tpy N as urea recently started in 1983 which makes the country independent of imports and gives it an export potential of about 30,000 tpy N, most of which will probably go to Australia. The plant is operated by the Petrochemical Corporation of New Zealand (Petrocorp).

New Zealand

Nitrogen Fertilizer Production, Consumption and Trade ('000 tons N)

	1979/80	1980/81	1981/82	1982/83	1983/84	1984/85
Production	-	-	-	5	48	55
Consumption	23	20	22	28	31	40
Exports	-	-	-	-	31	30
Imports	25	20	22	25	25	26

1.34 The plant at Kapuni is based on off-shore natural gas.

4. Other Developed Market Economies

The three countries in this economic classification are Israel, Japan and South America.

Israel

1.35 Israel has a relatively low demand for nitrogen fertilizers and consumption is only expected to rise above 50,000 tpy by the end of the decade. There is a small ammonia plant belonging to Fertilizers and Chemicals Limited at Haifa producing about 70,000 tpy N. Part of this production is used for the manufacture of potassium nitrate which Israel exports as a speciality fertilizer. Little growth is expected in the nitrogen industry in Israel.

Israel

Nitrogen Fertilizer Production, Consumption and Trade ('000 tons N)

	1979/80	1980/81	1981/82	1982/83	1983/84	1984/85
Production	76	62	62	74	82	71
Consumption	36	40	36	43	47	52
Exports	25	22	21	23	25	27
Imports	3	6	7	7	11	10

1.36 Israel has no reserves of natural gas and the plant at Haifa is based on naphtha.

Japan

1.37 Japan has a relatively high but stagnant demand for nitrogen both for fertilizers and industrial use. Although nitrogen demand for both domestic and export use has fluctuated between 1.4-1.6 million tpy over the last ten years, there has been no significant upward trend and little growth is expected in the future. Japan has been slowly losing its competitive position as a major exporter of nitrogen fertilizers, as its naphtha based ammonia industry has found it increasingly difficult to compete with gas based producers elsewhere in the world. Japanese ammonia production reached its peak in 1974/75 with about 3.3 million tons N but as part of an industry rationalization, capacity was afterwards reduced to about 2.8 million tons N. Under the Temporary Measures Law for the Structural Adjustment of Specific Industries adopted in June 1982 the Japanese ammonia industry must reduce its capacity by 20% to 2.24 million tons N by 1985/86. Even so, with a capacity of 2.24 million tons and a demand currently of only 1.9 million tons, due mainly to a fall in urea exports and caprolactam production, there will still be some excess capacity. Japan has closed many of its old small plants and in one situation, Ube Industries Ltd. have replaced a large naphtha based plant with a 1,000 tons per day (tpd) plant which came on-stream in 1984 using the Texaco coal gasification process. At the time it was claimed to reduce production costs by about 20% but the recent decline in Naphtha prices will reduce this advantage. There are eight producers of ammonia in Japan and the most important of these are Nissan Chem., Mitsui-Toatsu Chem., and Ube Ind. Ltd.

Japan

Nitrogen Fertilizer Production, Consumption and Trade ('000 tons N)

	1979/80	1980/81	1981/82	1982/83	1983/84	1984/85
Production	1,458	1,202	1,253	1,126	1,076	1,211
Consumption	777	614	643	683	701	697
Exports	701	580	470	333	227	310
Imports	35	42	48	73	80	72

1.38 Japan has no domestic resources of oil or gas and must depend on imports for feedstock for its ammonia industry. Some plants utilize coke oven gas but until the late 1970s most plants were based on naphtha. Following the increase in naphtha prices many plants converted to imported Liquified Petroleum Gas (LPG).

1.39 In 1982 however, due to the depressed economy, naphtha prices fell but at the same time the price of LPG remained stable due to lack of availability. This was caused by a reduced production of associated gas resulting from the decline in crude oil production. As a result of this, most of the main ammonia producers are considering multipurpose feedstock plants which would allow them to change from one feedstock to another as the market price changes. As already mentioned, Japan has recently commissioned a plant based on coal gasification.

South Africa

1.40 After Egypt, South Africa is the largest nitrogen market in continental Africa, accounting for about 25% of demand. Both consumption and production of nitrogen fertilizers rose steadily during the 1970s in South Africa, but in the period 1981-84 fertilizer consumption fell, mainly as a result of the drought and depressed agricultural markets. There is however a good potential for nitrogen fertilizer use in South Africa and in the longer-term, nitrogen fertilizer demand is expected to increase to about 600,000 tpy N by the end of the decade.

1.41 There are three major producers of ammonia in South Africa, African Explosives and Chemicals Ind. Ltd. (AECI), Fedmis and South Africa Coal, Oil and Gas (SASOL). A large coal based plant had been planned by Sentrachem, the mother company of Fedmis, to come on-stream in 1985, but this has now been cancelled. In order to meet future demand for ammonia that may have to come from imports, a large ammonia terminal has been erected at Richards Bay. One use of this terminal might be to receive ammonia from a new project now being considered in nearby Mozambique.

South Africa

Nitrogen Fertilizer Production, Consumption and Trade ('000 tons N)

	1979/80	1980/81	1981/82	1982/83	1983/84	1984/85
Production	433	444	447	409	397	482
Consumption	403	466	527	474	368	406
Exports	26	31	7	27	32	85
Imports	18	72	126	148	27	10

1.42 Following the recent moth-balling of the AECI plant at Umbogintwini which was based on naphtha/refinery gas, most of the operating capacity in South Africa is coal based.

B. Developing Market Economies

1. Africa

1.43 Within the FAO economic classification of countries, the major nitrogen consuming and/or producing countries South Africa, Libya, Egypt are dealt with in other sections. Of the remaining countries in Africa, the most important consumers are Algeria, Morocco, Nigeria, Zimbabwe and Zambia and the only significant producer is Algeria although there are plans for several large new ammonia plants in other countries.

Algeria

1.44 Algeria has a relatively small demand for nitrogen fertilizers of about 60,000 tpy N. Although Algeria has been attempting to establish a nitrogen industry based on its large reserves of natural gas it has been only partially successful and capacity first commissioned in 1969 never worked properly and is now being refurbished and will recommence operation in 1985/86. Because of these production delays, Algeria remained a net importer of nitrogen until 1981 when a new plant at Arzew came on-stream. (Algeria is now a net exporter of N because of ammonia exports but it still remains a net importer of nitrogen fertilizers.) An ammonia plant under construction at Annaba was due to come on stream in 1986. Asmidal, formerly Sonatrach, is considering a 1000 tpd ammonia plant at Skikda, to come on-stream by the end of the decade which would allow Algeria to become a major exporter of nitrogen fertilizers.

Algeria

Nitrogen Fertilizer Production, Consumption and Trade ('000 tons N)

	1979/80	1980/81	1981/82	1982/83	1983/84	1984/85
Production	21	24	24	51	47	65
Consumption	62	84	71	49	59	83
Exports	-	-	3	3	1	1
Imports	34	65	37	19	10	17

1.45 Algeria has large reserves of natural gas of which about 85% is non-associated. At one time a considerable amount of gas was flared but this has almost stopped. Algeria has been established as an exporter of liquified natural gas (LNG) for almost twenty years, although plans to develop the industry in the 1960s have not developed to the extent anticipated. Algeria will also export gas to Europe through the Trans-Mediterranean pipeline system some of which will go to ammonia production. Algeria has the potential to expand its production of natural gas and hence nitrogen fertilizer production in the future. It is well placed geographically to meet the growing needs of South Europe and other markets.

Morocco

1.46 Morocco currently consumes about 80,000 tons per year of nitrogen fertilizers most of which are imported. With increasing emphasis on agriculture, demand is expected to improve through the remainder of the decade. The country is a major producer and exporter of diammonium phosphate which is based on imported ammonia and so the country's needs of ammonia will increase significantly as Morocco increases its production of processed phosphates. Morocco has considered the erection of an ammonia plant at Jorf Losfar to meet its growing needs but this is only likely to materialize if natural gas reserves well above present levels can be found. Otherwise Moroccan imports of nitrogen as ammonia and fertilizers will probably increase to about 350,000 tpy N by the end of the decade.

Morocco

Nitrogen Fertilizer Production, Consumption and Trade ('000 tons N)

	1979/80	1980/81	1981/82	1982/83	1983/84	1984/85
Production	20	36	26	41	55	34
Consumption	101	123	81	92	112	104
Exports	1	1	3	15	22	5
Imports	87	90	46	98	68	97

1.47 Morocco's reserves of gas are relatively small and insufficient to establish a nitrogen fertilizer industry, although further exploration and discoveries could well change this situation.

Nigeria

1.48 Nigeria has a tremendous need to use more nitrogen fertilizer to feed its growing population and restore a once prosperous agricultural export business. Nitrogen fertilizer consumption grew rapidly up to 1981 but then stagnated with the major economic problems of 1982. In order to assist the Government to protect its agricultural programs, in 1983/84 the World Bank financed a project to import two million tons of fertilizer. Imports of fertilizer will continue to be a major problem for Nigeria until the first major nitrogen fertilizer complex is completed in 1988/89. This project is a joint venture between the Government of Nigeria and Kellogg of the USA and will produce ammonia, urea and NPK compounds.

Nigeria

Nitrogen Fertilizer Production, Consumption and Trade ('000 tons N)

	1979/80	1980/81	1981/82	1982/83	1983/84	1984/85
Production	-	-	-	-	-	-
Consumption	68	92	97	92	123	131
Exports	-	-	-	-	-	-
Imports	62	92	97	92	123	131

1.49 Nigeria has very large reserves of natural gas estimated at 1,370 BCM spread over many fields in the two main producing basins, Niger Delta and Anambra. About one third of this is associated gas. Until recently almost all (about 90%) of the natural gas was flared, with small amounts being used in the Port Harcourt refining and other industries. Under current legislation flaring of gas must now cease. There is thus ample available feedstock to develop a nitrogen fertilizer industry to meet current domestic needs and future export requirements, although the problems of implementing new projects will still remain a major constraint.

Zambia

1.50 Zambia is the third largest fertilizer consumer in East Africa after Zimbabwe and Sudan and nitrogen fertilizer consumption grew steadily during 1975-1983 with an average yearly increase of 8.7% from 33,000 tons to 58,000 tons. Since then, however, it has declined to 44,000 tons in 1983 due to the depressed agricultural situation. About 25% of total nitrogen demand in Zambia is used for the manufacture of explosives for the copper industry.

1.51 Nitrogen Chemicals of Zambia Ltd. (NCZ), is the sole producer of nitrogen fertilizer in Zambia. The company was formed in 1967 to construct and operate a coal based ammonium nitrate plant at Kafue (30 km from Lusaka) mainly to produce explosive grade ammonium nitrate acid for the mining industry. The scope of the project was expanded to include ammonium nitrate for fertilizers.

The initial ammonia capacity was about 25,000 tpy N but this was increased by another plant with a capacity of 57,000 tpy N in 1981. The supply capability of the plants has so far remained well below design capacity. Consequently Zambia has had to rely heavily on imports to meet its nitrogen requirements.

Zambia

Nitrogen Fertilizer Production, Consumption and Trade ('000 tons N)

	1979/80	1980/81	1981/82	1982/83	1983/84	1984/85
Production	8	7	4	9	16	15
Consumption	38	53	66	58	44	37
Exports	-	-	-	-	-	-
Imports	35	31	53	58	44	61

1.52 Ammonia is produced in Zambia by gasification of locally produced coal.

Zimbabwe

1.53 Demand for nitrogen fertilizers grew rapidly in Zimbabwe through the 1960's but the economic and political problems of the 1970s constrained growth. Demand took off again after the establishment of independence at the end of the 1970s. Ammonia production first started in Zimbabwe in 1972 when a small ammonia plant was commissioned by Sable Chemical industries at Que Que using electrolysis of water as a source of hydrogen. Previously ammonium nitrate had been produced using imported ammonia.

Zimbabwe

Nitrogen Fertilizer Production, Consumption and Trade ('000 tons N)

	1979/80	1980/81	1981/82	1982/83	1983/84	1984/85
Production	59	72	70	84	71	69
Consumption	59	93	98	73	81	72
Exports	1	-	-	-	-	-
Imports	4	21	37	42	14	10

1.54 The production of ammonia in Zimbabwe is from hydrogen produced from hydroelectric power.

Other African Countries

1.55 Although both consumption and production of nitrogen fertilizers at the present time are rather small, many countries in this region have resources of natural gas and in some cases hydroelectric power which could form the basis of a nitrogen fertilizer industry and these potential production sites are briefly outlined below.

Angola: This country produces oil from offshore fields from which gas is currently flared. There are also offshore non-associated natural gas reserves in the Congo River Basin. Angola is believed to be a good prospect for future discoveries of oil and gas and SONANGOL the national oil company is already investigating the prospects of building an ammonia plant at Quinsuquenna.

Cameroon: Cameroon has a significant and growing demand for nitrogen fertilizers and to meet this need a complex fertilizer plant was commissioned in 1977 at Bonaberi. Due to both technical and economic problems this plant has been idle most of the time since and the country depends largely on imports to meet its needs. At the end of the 1970s, after limited drilling, there were hopes of very large gas finds but estimates have now been reduced. Nevertheless, there are ample reserves on which to base a nitrogen fertilizer industry, both to serve a growing domestic market and also for export.

Madagascar: Although nitrogen consumption has been relatively stagnant at about 6000 tpy N, consumption is expected to rise steadily in the future following the commissioning of a small ammonia/urea complex which will produce about 30,000 tpy N as urea using naphtha as feedstock.

Cote d'Ivoire: Nitrogen fertilizer demand has varied between 10,000-20,000 tpy N. There is a mixed fertilizer granulation plant at Abidjan but all nitrogen intermediates are imported. The Government is looking at the possibilities of utilizing natural gas from the offshore oil fields for ammonia production.

Mozambique: Nitrogen fertilizer demand in Mozambique had risen to about 20,000 tons of N by 1983 but in 1984/85 it fell to 3,000 tons all of which was imported. There is a small ammonium sulphate plant which is based on imported ammonia. Natural gas reserves are located in the Pande gas field which could support a nitrogen fertilizer industry and a feasibility study has been carried out for a 1,000 tpd ammonia plant at Inhasso. Increasing demand for ammonia in South Africa could provide a potential market for this plant.

Tanzania: This country consumed about 15,000 tpy nitrogen in 1984, a decline of nearly 10,000 tpy since 1979, due mainly to the shortage of foreign exchange. A large ammonia/urea complex is planned to come on-stream at the end of the decade and although this is intended mainly for export it should also boost local demand. The plant situated at Kilamco will be based on gas from the Songo Songo field.

Zaire: A project has been proposed to build a 1,000 tpd ammonia plant in Zaire. The project, which would be based on hydrogen produced by electrolysis is being promoted by Canadian interests.

2. Latin America

1.56 Latin America which includes Central America, the Caribbean area and Central America, is one of the most important developing regions both with regard to the production and use of nitrogen fertilizers. Many countries in the region such as Trinidad, Mexico, Venezuela, etc. are already exploiting their large natural gas resources. Major consumers are Brazil and Mexico and, with a growing population and the potential to increase its agricultural production, the region is expected to have one of the highest growth rates in nitrogen demand in the future.

Argentina

1.57 Nitrogen fertilizer demand in the last two years has grown rapidly assisted by an enterprising agricultural program which supplies fertilizer to farmers in exchange for wheat production and which in 1984/85 showed a significant increase over that in 1983/84. There are three small nitrogen fertilizer plants in Argentina producing about 30,000 tpy nitrogen, mainly in the form of urea and ammonium sulphate. With the prospects of a rapidly growing demand for nitrogen fertilizers to increase agricultural production and with the availability of cheap natural gas, the Government has agreed to the construction of an ammonia/urea complex. It is anticipated that part of the production at least initially will go to the export market. The Government may also approve the construction of one or two small ammonia plants. Ownership of these plants will be in the private sector probably with overseas participation. In the short-term Argentina will depend on imports of urea to meet its growing demand but the increasing revenues from grain exports should more than justify the cost of fertilizer imports.

Argentina

Nitrogen Fertilizer Production, Consumption and Trade ('000 tons N)

	1979/80	1980/81	1981/82	1982/83	1983/84	1984/85
Production	25	30	25	29	32	34
Consumption	59	63	51	53	65	98
Exports	1	-	-	-	-	-
Imports	41	35	26	24	32	64

1.58 Argentina has large reserves of natural gas situated mainly in the southern tip of the country and the geology of the deposit indicates a potential for future discoveries. About 25% of total gas production is now being flared but the extension of gas pipelines to areas where gas can be used will increase the potential to use this gas for ammonia production.

Brazil

1.59 Nitrogen fertilizer demand grew rapidly during the 1960s due to favorable agricultural credit terms. Despite a setback in the mid-1970s, caused by high raw material prices, growth continued during the 1970s. In 1980/81 nitrogen fertilizer consumption exceeded 900,000 tons N but since then, due to adverse economic conditions and changes in both agricultural prices and credit, consumption declined by about 40% in the next three years. In 1984/85 with more favorable economic conditions in the country, nitrogen fertilizer demand increased by almost 50% and made up much of the loss of the early 1980s.

1.60 Brazil's nitrogen fertilizer industry has undergone considerable expansion in the last few years and now has an ammonia capacity of about one million tons N, due mainly to the commissioning of three large scale ammonia/urea plants by Petrobras at Camacari in 1978, Araucaria in 1981 and Laranjeiras in 1982. Although there have been plans for several other major projects to meet the demand anticipated based on growth rates of the late 1970s, with the decline in demand in the 1980s, Brazil for a short while actually became an exporter of urea and most of the plans for expansion were cancelled. With an improved outlook for nitrogen demand Brazil's net import needs are increasing and consideration is once again being given to a new plant based on gas in Campos.

Brazil

Nitrogen Fertilizer Production, Consumption and Trade ('000 tons N)

	1979/80	1980/81	1981/82	1982/83	1983/84	1984/85
Production	284	384	350	397	533	670
Consumption	779	905	668	642	553	814
Exports	2	1	1	1	73	10
Imports	496	522	319	347	104	153

1.61 About half of Brazil's nitrogen industry is based on natural gas with the remainder using an assortment of other types of feedstock. Brazil has some associated gas available at the Campos oil field near Rio de Janeiro which is being considered as a feedstock for a new ammonia/urea complex.

Chile

1.62 Nitrogen fertilizer consumption in Chile recovered in 1984/85 after a fall in the early 1980s and has now exceeded its peak level of 1974. About half of nitrogen demand is satisfied from domestically produced nitrates and the remainder comes from imported urea. Chile exports about 70,000 tpy N as sodium nitrate and imports about 20,000 tpy ammonia for the production of explosive grade ammonium nitrate. Consideration is now being given to the construction of a large export based ammonia/urea plant based on natural gas.

Chile

Nitrogen Fertilizer Production, Consumption and Trade ('000 tons N)

	1979/80	1980/81	1981/82	1982/83	1983/84	1984/85
Production	94	101	93	92	98	113
Consumption	51	51	48	48	65	86
Exports	71	78	68	58	67	59
Imports	28	28	35	9	28	43

1.63 Chile has significant reserves of natural gas, mainly in the remotely located Magallanes area. Several projects are being considered to exploit the gas including LNG, methanol and one ammonia project.

Colombia

1.64 Colombia is one of the largest users of nitrogen fertilizers in Latin America, consuming nearly 160,000 tons N in 1983/84. Consumption is expected to increase at about 3% py through the 1980s. About one third of this consumption is supplied from domestic production in small plants at Barranquilla, Barrancabermeja, and Cartagena. The Colombian Government with technical assistance from International Agencies has been examining the prospects for a large ammonia/urea complex utilizing natural gas to supply both domestic and export markets. One possibility considered is a plant at Palamino Guajira on the Atlantic Coast.

Colombia

Nitrogen Fertilizer Production, Consumption and Trade ('000 tons N)

	1979/80	1980/81	1981/82	1982/83	1983/84	1984/85
Production	57	42	42	48	54	60
Consumption	151	143	155	150	157	181
Exports	2	3	2	-	3	-
Imports	87	103	89	101	109	126

1.65 Colombia's small existing ammonia plants are based on domestic natural gas. Most proven reserves are located in the Guajira area of the northwest, mainly in the Chuchupa field and this would be the source of gas for the projected ammonia/urea complex.

Cuba

1.66 Although a centrally planned economy, Cuba still comes within the FAO classification for Latin America. Cuba is one of the largest nitrogen fertilizer consumers in the region and growth in demand is expected to continue at about 5% py through 1990. Although Cuba commissioned two ammonia plants in 1976 which should have been adequate to meet its needs, the plants have not operated well, particularly the urea plant at Cienfuegas. Consequently about two thirds of nitrogen fertilizer consumed are imported.

Cuba

Nitrogen Fertilizer Production, Consumption and Trade ('000 tons N)

	1979/80	1980/81	1981/82	1982/83	1983/84	1984/85
Production	131	112	142	96	74	169
Consumption	280	267	315	275	249	294
Exports	-	-	-	-	-	-
Imports	142	155	173	181	191	196

1.67 Cuba has no resources of natural gas and the ammonia plant at Neuvitx uses fuel oil and the plant at Cienfuegas uses naphtha.

Ecuador

1.68 Nitrogen fertilizer consumption in Ecuador fluctuated between 40,000 and 60,000 tpy of N from 1975 to 1980, but fell during the early 1980's to 33,000 tpy in 1982/83. Consumption recovered to about 50,000 tpy the following year. There is a small ammonia storage facility and ammonium sulphate plant at Guayaquil but production has not exceeded 3,000 tpy N in the last five years or so. From time to time there are reports of an ammonia/urea project being developed based on off-shore gas but so far no firm proposals have materialized. The proposed site for the plant is Pasorja.

Ecuador

Nitrogen Fertilizer Production, Consumption and Trade ('000 tons N)

	1979/80	1980/81	1981/82	1982/83	1983/84	1984/85
Production	2	2	2	3	-	-
Consumption	43	41	35	33	47	43
Exports	-	-	-	-	-	-
Imports	46	38	32	30	47	43

1.69 Ecuador has significant reserves of natural gas which may increase as the Amistad field is further explored. The gas from Amistad will be pumped to the shore - 70 miles away, for exploitation.

Mexico

1.70 With more than one million tons of nitrogen fertilizer consumption, Mexico is the largest consumer in the region. With large reserves of natural gas it is also the largest producer of nitrogen fertilizers in the region. In 1979, at its three major production sites of Camargo, Salamanca and Cosaleacacque, Mexico already had a capacity of 1.7 million tons of N as ammonia, and by the end of the decade, with several new plants planned to come on-stream, this capacity should almost double. Although financial constraints a few years ago necessitated a slow-down in the plans to commission new nitrogen fertilizer plants, the projected tight situation for nitrogen fertilizers at the end of the decade once again activated the postponed projects. However, the recent fall in oil prices and hence revenues to the country may once again slow down development.

1.71 Mexico has been a continuous importer of nitrogen fertilizers, mainly as urea but it has also at the same time had a surplus ammonia capacity which was not matched by downstream capacity. As a result of this, Mexico exported large quantities of ammonia during a period when it was importing nitrogen fertilizers.

Ammonia Exports from Mexico ('000 tons N)

1980	1981	1982	1983	1984
584	642	682	609	406

These exports are expected to decline as new downstream facilities are completed at Lazaro Cardenas.

Mexico

Nitrogen Fertilizer Production, Consumption and Trade ('000 tons N)

	1979/80	1980/81	1981/82	1982/83	1983/84	1984/85
Production	642	739	877	1,067	1,044	1,188
Consumption	826	904	1,112	1,255	1,088	1,193
Exports	-	-	-	3	88	58
Imports	185	172	237	190	53	52

1.72 Mexico has very large reserves of natural gas most of which are found in association with crude oil. The utilization of natural gas has always had a high priority and although in 1975 about 20% of natural gas was flared this has now been reduced to less than 5%. About 30% of gas production goes into the manufacture of petrochemicals including ammonia and there are plans to increase the production facilities for both ammonia and methanol. The rate at which Mexico further develops its nitrogen industry will depend mainly on its ability to finance the projects, as both domestic and export markets for ammonia seem likely to increase in the future. Production and use of gas is controlled by the state-owned oil and gas company PEMEX, which also produces ammonia. Urea is produced by the State Fertilizer Company, FERTIMEX.

Trinidad & Tobago

1.73 Although chemical fertilizer demand in this country is very small, Trinidad is a major producer of ammonia and nitrogen fertilizers. With its relatively good location for the major nitrogen markets of the USA, South America and Europe, and a good supply of natural gas, the country is well placed to expand its nitrogen fertilizer industry. One new plant has been approved (Tringin II) and others are being considered. The recent fall in energy prices may delay some of these plans.

1.74 There are already four large ammonia plants with a total capacity of more than one million tons operating in Trinidad. The first of these plants, FEDCHEM, a subsidiary of W. R. Grace, USA, commenced operation at Point Lisas in 1959. In 1978 a second plant - TRINGIN (49% Grace, 51% Government) came on-stream followed by two more in 1981/82 - FERTRIN (49% Amoco, 51% Government). TRINGEN plans a new large plant to come on-stream 1988.

Ammonia Exports from Trinidad ('000 tons N)

1980	1981	1982	1983	1984
475	440	733	981	966

Trinidad

Nitrogen Fertilizer Production, Consumption and Trade ('000 tons N)

	1979/80	1980/81	1981/82	1982/83	1983/84	1984/85
Production	44	41	23	29	20	67
Consumption	5	4	3	3	4	5
Exports	40	37	20	28	19	45
Imports	-	-	-	-	2	2

1.75 Reserves of gas are estimated to be more than 400 BCM, mainly as associated gas and with high methane content. (About 25% of current gas production is used for ammonia manufacture.)

Venezuela

1.76 After growing steadily from 65,000 tons N in 1975/76 to 113,000 tons in 1980/81, fertilizer consumption slumped back to its 1975/76 level in

1981/82. This fall in consumption was due mainly to a reduction in fertilizer subsidies. Demand has improved in the last two years and has reached its previous high level. Although the nitrogen fertilizer industry in Venezuela has a relatively long history, the performance of the production operations has not been good and generally plants have worked at well below half their rated capacity. Although rehabilitation schemes are being implemented, production is still low but should improve significantly by 1986 when the improvements are completed. Venezuela is planning to increase its nitrogen capacity during the next decade with an ammonia plant at El Tablazo and an ammonia/urea complex at Caripito.

Ammonia Exports from Venezuela ('000 tons N)

1980	1981	1982	1983	1984
128	169	193	100	166

Venezuela

Nitrogen Fertilizer Production, Consumption and Trade ('000 tons N)

	1979/80	1980/81	1981/82	1982/83	1983/84	1984/85
Production	102	144	165	208	206	255
Consumption	97	113	69	70	73	123
Exports	74	70	113	105	110	138
Imports	55	53	54	14	2	12

1.77 Venezuela has large reserves of natural gas of more than 1,500 BCM, second only to Mexico in Latin America. Much of the gas is associated with oil production and is being reinjected, with only a small amount flared. New discoveries of non-associated gas have been made in the North Paria Basin, offshore, near the mouth of the Orinoco River. Venezuela has the material base to develop its nitrogen production facilities in the future and is planning several new nitrogen projects.

Other Countries in Latin America

1.78 Although the major countries with regard to fertilizer demand or production and potential nitrogen fertilizer production have been outlined, there are several other countries where nitrogen fertilizer use is significant and important to the agricultural needs of the country and as an import market for the region.

Costa Rica: Nitrogen fertilizer demand in Costa Rica has increased in the last ten years from about 30,000 tpy of N to 50,600 tpy of N in 1984/85. There is no basic ammonia production but about 50,000 tpy ammonia is imported and converted into ammonium sulphate and ammonium nitrate for use as components of mixed fertilizers. The plant is operated by FERTICA-SA-Guanomex at Puntarenus.

El Salvador: Nitrogen fertilizer demand had reached about 84,000 tpy N by 1979 but unfortunately due to the economic and political problems in the country, demand has now fallen to half that level. El Salvador has no production of ammonia although small quantities have been imported by FERTICA to be converted into mixed fertilizers at its plant at Acajutla. This plant however has not operated recently.

Guatemala: Nitrogen fertilizer consumption has increased steadily and almost doubled in the last ten years to reach 55,500 tons in 1984/85. There is no basic ammonia capacity in the country but a small quantity of ammonia is imported for the production of compound fertilizers in the FERTICA SA - Guanomex plant at Cuidad Tecunuman.

<u>Peru</u>: Nitrogen fertilizer consumption in Peru had reached a figure in excess of 100,000 tons for several years during the 1970s but over the last two or three years has fallen to half that level due to balance of payment problems. There are two small nitrogen fertilizer complexes, one at Cuzco and the other at Callao, as well as the ammonia urea complex of Talara. The Talara plant uses natural gas whereas the small ammonia plants at Guzco and Callao are based on electrolysis and fuel oil respectively.

3. Near East

1.79 The main consuming countries are Turkey, Egypt and Iran, which account for about 75% of the consumption in the region. The area is rich in natural gas and several countries are major producers of ammonia so that overall the region is a net exporter of nitrogen.

Turkey

1.80 Demand for nitrogen fertilizers has almost doubled over the last ten years, despite a slight check in 1981/82, reaching 995,000 tons in 1984/85. The Turkish Government had subsidized fertilizers during the 1970s to improve agricultural production but reduced these subsidies during the economic crisis of 1980 - hence the fall in consumption in the following year. With an improving economic situation and guaranteed agricultural prices, demand has now started to recover. Turkey has still one of the lowest fertilizer application rates compared with developed countries and so there is still a considerable potential for further growth.

1.81 Although the nominal nitrogen capacity should be adequate to meet demand, the industry has suffered from a mixture of technical problems and shortage of raw materials which has inhibited manufacture. In order to help modernize and restructure the industry the World Bank is providing financial assistance to Turkey.

1.82 As the downstream facilities are improved, Turkey is replacing its imports of finished nitrogen fertilizers by ammonia, and imports have increased significantly making Turkey one of the largest importers of ammonia, a situation that is likely to continue in the future.

Ammonia Imports to Turkey ('000 tons N)

1980	1981	1982	1983	1984
230	391	299	548	471

Turkey

Nitrogen Fertilizer Production, Consumption and Trade ('000 tons N)

	1979/80	1980/81	1981/82	1982/83	1983/84	1984/85
Production	340	672	718	723	783	759
Consumption	685	807	799	863	1,021	955
Exports	-	-	-	2	13	-
Imports	450	331	128	118	257	294

1.83 Turkey has only limited reserves of natural gas and existing plants use an assortment of other feedstocks such as naphtha, coke oven gas and lignite. Turkey is likely to depend to an increasing extent on imported ammonia, although it is also reported that Turkey is considering large increased imports of Soviet gas which could be used as a feedstock for ammonia production.

Egypt

1.84 Although Egypt is classified as part of the Near East, in fact it has the largest nitrogen fertilizer consumption in continental Africa. Egypt's nitrogen fertilizer use has grown steadily over the last few years and is expected to increase at about 5% per year through the remainder of the decade. Until recently, Egypt depended on imports to meet about 40% of its consumption but following the commissioning of the large ammonia/urea complex, Talka II, in 1983/84, the country was actually a small net exporter of nitrogen. Egypt will however need new capacity soon, both to meet increasing demand, and also to replace the high cost ammonia plant at Kima which produces ammonia using electrolysis.

Egypt

Nitrogen Fertilizer Production, Consumption and Trade ('000 tons N)

	1979/80	1980/81	1981/82	1982/83	1983/84	1984/85
Production	264	400	482	626	646	680
Consumption	500	554	585	668	722	649
Exports	-	1	2	1	82	138
Imports	208	164	163	36	26	14

1.85 Current capacity is based on natural gas, electricity, refinery gas and coke oven gas. Although Egypt has the longer term potential to produce more gas, the current availability and usage of gas is relatively small. The Government has not encouraged export-based plants until availability of gas is increased and new developments will be basically for domestic use.

Iran

1.86 Iran has one of the largest markets for nitrogen fertilizers in the Near East and despite the current hostilities in the area, fertilizer consumption has increased markedly in the last two or three years. In 1983/84 nitrogen fertilizer consumption reached nearly 600,000 tons N and almost all of this was imported. In 1984/85, consumption fell to less than 500,000 tons N.

1.87 Before the war with Iraq, Iran had a small ammonia plant in operation at Shiraz and had just commissioned two large plants at Bandar Khomeni. A further plant was under construction at Shiraz and these plants would have in aggregate given Iran an ammonia capacity of about 850,000 tons of nitrogen, which would have easily met Iran's short term needs. As a result of the war, the two plants at Bandar Khomeni were closed and work on the new plant at Shiraz was much delayed though it was expected to start up in 1985/86.

1.88 In the present situation, the future for the nitrogen industry in Iran is very uncertain. The country has one of the largest reserves of natural gas in the world and with other fertilizer raw materials such as sulphur and large potential domestic demand and export potential the country is very well placed to develop a major fertilizer industry.

1.89 It will obviously take some time for the country to recover from the ravages of the war, even if peace comes soon. The extent of the damage to the plants at Bandar Khomeni is unknown but it seems likely that refurbishing of the plants will be necessary before full production is achieved and throughout the remainder of the 1980s Iran will continue to be a large and growing importer of nitrogen fertilizers.

Iran

Nitrogen Fertilizer Production, Consumption and Trade ('000 tons N)

	1979/80	1980/81	1981/82	1982/83	1983/84	1984/85
Production	9	71	21	22	24	14
Consumption	237	275	387	493	597	487
Exports	-	-	-	-	-	-
Imports	157	212	393	472	572	472

1.90 Iran has the second largest reserves of gas in the world (after the USSR). About 40% of the reserves are associated with the oil fields in Khuzestan and about 50% is in large non-associated fields, onshore near Kangom and extending into the Persian Gulf. Many of the operations and plans to develop these gas reserves have been suspended or delayed because of the war. About 50% of gas currently produced is flared and, when peace comes once again, the country is a major candidate for developing a gas based nitrogen fertilizer industry.

Kuwait

1.91 Kuwait's demand for nitrogen fertilizers is almost insignificant but the country is one of the major producers and exporters of nitrogen fertilizers. The state owned company Petrochemicals Industry Co. (PIC) operates the nitrogen plants at Shuaiba which, after 1985, will have a capacity of more than 500,000 tons of N. Because of the proximity of Kuwait to the current hostilities in the area and the fear of damage to the ammonia storage facilities, there has been reduced production in the last few years.

1.92 Apart from its interests in Kuwait, the Government has been interested in a number of ventures overseas particularly in collaboration with Tunisia. It has a stake in the Tunisian fertilizer industry and will also participate with Tunisia in a joint project with the Chinese Government in China for the manufacture of diammonium phosphate.

Kuwait

Nitrogen Fertilizer Production, Consumption and Trade ('000 tons N)

	1979/80	1980/81	1981/82	1982/83	1983/84	1984/85
Production	307	218	183	234	224	281
Consumption	-	-	-	1	1	-
Exports	291	226	176	196	241	271
Imports	-	-	-	-	-	-

1.93 Nearly all reserves of gas are associated gas and gas production is therefore limited by oil production. At current oil production rates all gas is committed. Although exploration is underway to establish other gas sources, it seems unlikely that there will be significant further development of Kuwait's nitrogen industry throughout the 1980s.

Saudi Arabia

1.94 Nitrogen fertilizer demand in Saudi Arabia is still relatively small although Government subsidies to improve agricultural production have boosted fertilizer consumption from virtually zero to 130,000 tpy N in the last ten years. The country has now an established nitrogen fertilizer industry and there are plans to expand this further. The first plant of the Saudi Arabia Fertilizer Co. (SAFCO) was commissioned at Damman at the end of 1969. In 1983 Saudi Arabia commissioned its second plant on the industrial estate of Al Jubail, a joint venture between the Taiwan Fertilizer Co. and Saudi Basic Industries Corp. (SABIC). In addition, SABIC and SAFCO are planning a large ammonia plant at Al Jubail which should come on-stream 1988/89.

Saudi Arabia

Nitrogen Fertilizer Production, Consumption and Trade ('000 tons N)

	1979/80	1980/81	1981/82	1982/83	1983/84	1984/85
Production	137	152	159	161	266	397
Consumption	16	25	41	59	113	130
Exports	134	131	133	138	167	326
Imports	4	10	17	20	48	47

1.95 Saudi Arabia has large reserves of natural gas, most of this as associated gas. In the past a large amount of gas was flared, but taking into account the increased commitments for this gas and reduced oil production there will be a need for new gas supplies by the end of the 1980s. This situation has increased the exploration for non-associated gas which is believed to lie under the large oil fields.

Other Near East Countries

1.96 Some brief comments on the nitrogen industry in other countries of the region follow.

Abu Dhabi: At the present time most of the natural gas reserves are associated, and at the current level of oil production, are committed. One use of this gas is for an ammonia plant at Ruwais which came on-stream 1983/84. This is a joint venture between the Abu Dhabi National Oil Company (ADNOC) and Compagnie Francais des Petroles (CFP).

Afghanistan: Consumption rose steadily to about 50,000 tons in 1978/79 but fell sharply following the Soviet invasion. A small ammonia/urea complex with a capacity of about 58,000 tpy N provides most of the country's needs.

Bahrain: Bahrain already has a number of industries based on its natural gas resources and most of its proven reserves are now committed. These include the Gulf Petrochemical Industries (GPIC) ammonia plant at Sitra, due on-stream 1985/86. This is a venture between Bahrain National Oil Co. Kuwait's PIC and SABIC of Saudi Arabia. The product will be exported.

Iraq: Like Iran, the nitrogen industry of Iraq has suffered because of the conflict between the two countries. The plants at Basrah and Khor-Al-Zubair amounting to about 0.5 million tons have been damaged and closed down. A small plant at Al Quam commissioned in 1983 is believed to be operating. Recently, it has been announced that a 1,670 tpd ammonia/urea complex will be built at Baiji.

Jordan: There is little agricultural demand for nitrogen in Jordan and no production. As a major producer of diammonium phosphate for exports, Jordan import needs for ammonia have increased to about 60,000 tons and are expected to increase to more than 100,000 tpy N in the future.

Qatar: A nitrogen industry was established in 1973 to utilize the large quantities of associated gas from the Dukhan field. A second plant was commissioned in 1979 raising the capacity to nearly 0.5 million tpy N. Norsk Hydro (Norway) has a 25% stake in the company, operates the plant and markets the product.

Sudan: In 1981 and 1982 nitrogen fertilizer consumption exceeded 80,000 tpy but due to balance of payment problems this has fallen considerably in the last year or so. A plant designed to produce about 100,000 tpy N has been delayed due to changes in policy regarding the availability of feedstock from domestic oil production.

Syria: Nitrogen fertilizer consumption in Syria has grown steadily and increased threefold in the last decade to about 127,000 tons in 1984/85. In 1982 a new nitrogen fertilizer plant came on-stream at Homs to produce 1,000 tpd ammonia. Until the operation of this plant is established fully, Syria will continue to import but probably after about 1986 the country should be a net exporter of nitrogen. Although the plant was initially based on naphtha it will be converted to natural gas when the pipeline from the Jubaisa and Suwerda fields is completed.

4. Far East

1.97 The countries in the Far East have, in the past two decades, shown a spectacular growth in both nitrogen fertilizer consumption and production. The major countries are India, Indonesia, Pakistan and Bangladesh. With a large and growing population, each of these countries has needed to increase its fertilizer production accordingly. The fact that natural gas is readily available in the area has also accelerated the growth of nitrogen fertilizer production.

Bangladesh

1.98 Bangladesh has maintained one of the highest nitrogen fertilizer demand growth rates in Asia, from 147,000 tons in 1975/76 to 340,000 tons N in 1984/85, an average yearly growth rate of about 10%. Demand is expected to slowdown to an average yearly growth-rate of about 6% over the next decade. Bangladesh has about 0.5 million tons of nitrogen capacity in operation. Up to 1981/1982, domestic production of nitrogen fertilizer was supplied by two plants at Fenchuganj and Ghorasal with total capacities of 230,000 tons per year of nitrogen, although actual production was considerably lower. At the end of 1981 a large plant was commissioned in Ashuganj with a capacity of 270,000 tons N per year as urea. A small urea plant has recently been commissioned at Ghorasal, built with Chinese aid, and a further plant will be brought on-stream at Chittagong, to bring nitrogen capacity, mainly in the form of urea, up to 826,000 tons per year by the end of the decade. These plants are owned by the government and operated by Bangladesh Chemical Industries Corporation (BCIC). A further plant is planned at Chittagong, which will be export-based initially.

1.99 Bangladesh has traditionally been a net importer of nitrogen fertilizers although imports have decreased as new capacity has come on stream.

Bangladesh

Nitrogen Fertilizer Production, Consumption and Trade ('000 tons N)

	1979/80	1980/81	1981/82	1982/83	1983/84	1984/85
Production	168	160	93	237	377	343
Consumption	260	268	252	306	356	399
Exports	-	19	4	33	26	12
Imports	141	39	124	35	67	81

1.100 All fertilizer production in Bangladesh is based on local natural gas. Recoverable gas reserves in Bangladesh are distributed among twelve fields. Of these twelve, only four have been developed for production. There are ample gas supplies to justify further nitrogen fertilizer capacity, both for domestic use and exports.

Burma

1.101 Over the last decade, consumption of nitrogen fertilizers in Burma has increased an average of about 12% per year. Most of the country's nitrogen fertilizer consumption is as urea and about two-thirds of this is produced domestically and the remainder is imported. Burma has two small urea plants, one located at Sale and the other at Kyunchaung with a combined capacity of about 66,000 tons of nitrogen. Production in recent years has achieved about 85% of the rated capacity. Expansion work at Sale has just been completed and will add another 49,000 tons N capacity. A third urea plant, with a capacity of 98,000 tons N per year was scheduled to be completed by the end of 1985. With a total capacity of 213,000 tons N by that time that plant comes on stream, Burma will be self-sufficient in nitrogen.

Burma

Nitrogen Fertilizer Production, Consumption and Trade ('000 tons N)

	1979/80	1980/81	1981/82	1982/83	1983/84	1984/85
Production	62	60	60	51	60	86
Consumption	70	67	93	115	115	127
Exports	-	-	-	-	-	-
Imports	4	25	46	67	55	59

1.102 Estimated natural gas reserves are about 200 BCM and with limited needs for the domestic market, Burma has the potential to build export-based ammonia/urea plants. Feasibility studies are currently underway for both export-based methanol and urea plants.

India

1.103 Following its independence, the first three Five-Year Plans during the period 1951/52 to 1965/66 laid special emphasis on increasing agricultural production. Most of the increase during this period came from increased net acreage under cultivation and it was not until the second half of the 1960s and throughout the 1970s that increased fertilizer consumption, used with new high yielding varieties, made its major impact. Between 1969/70 and 1984/85 fertilizer consumption increased from 1.36 million tons to 5.3 million tons-an average annual increase of nearly 10% per year. Nitrogen fertilizer consumption is expected to grow at about 6% per year throughout the remainder of the 1980s. In order to meet this rapidly increasing demand for nitrogen fertilizers India has made tremendous progress in developing nitrogen fertilizer production and with 35 ammonia plants in operation is ranked as the world's fourth largest nitrogen fertilizer producer. There are three gas-based nitrogenous fertilizer plants under construction scheduled to be completed by 1985/86 with a combined rated capacity of 1.5 million tons of N. In addition, eight more gas-based nitrogenous fertilizer plants, and one each based on naphtha and coal are in the planning stage. The target is to raise nitrogen capacity to a level of 9.9 million tons by 1989/90. In 1984, with a total installed capacity just over 5 million tons of nitrogen, 57% was in the public sector, 33% in the private sector and about 10% run by cooperatives.

1.104 Because India's consumption has steadily outpaced its ability to supply from domestic sources, the country has been a major importer of chemical nitrogen intermediates and finished fertilizers. The main import has been urea, although anhydrous ammonia is also imported. Fertilizer Association of India (FAI) are projecting that by the end of the decade, India's production will meet demand. This depends, however, on all new plants coming on-stream as planned. If the balance is to be maintained, further new plants will be required to come on-stream in the early 1990s.

India

Nitrogen Fertilizer Production, Consumption and Trade ('000 tons N)

	1979/80	1980/81	1981/82	1982/83	1983/84	1984/85
Production	2,224	2,164	3,143	3,430	3,487	3,917
Consumption	3,444	3,522	3,882	4,043	4,627	5,333
Exports	-	-	-	-	-	-
Imports	1,295	1,510	1,055	424	656	2,009

1.105 A wide variety of feedstocks is used to manufacture ammonia, such as natural gas, naphtha, coal, coke oven gas and hydrogen from water electrolysis. The majority of existing plants are based on naphtha, while most of the new plants at the planning stage are based on natural gas. In 1983,46.4% of installed capacity was based on naphtha, 13.9% on natural gas, 24.9% on fuel oil, 8.6% on coal and the small remainder from a variety of other sources. Proven reserves of natural gas in India are estimated at 475 BCM onshore and offshore. Most of this gas is located in non-associated South Bassein fields, the associated Bombay High field and in the Assam area. It is believed that the ultimate gas reserves will be much higher. Of the 11 new fertilizer plants planned or under construction based on natural gas, 10 will use Bombay high gas. These plants will be situated in the vicinity of the market and will be fed by gas through a pipeline from the coast.

Indonesia

1.106 There has been a rapid increase in fertilizer demand in Indonesia in recent years, due mainly to the favorable relationship set by the Government between the price of nitrogen fertilizers and rice. Between 1975/76 and 1984/85, nitrogen fertilizer consumption has increased more than threefold from 342,000 to 1,285,000 tons. Undoubtedly, one of the main reasons for this large increase in fertilizer use has been the rapid and successful development of the Indonesian nitrogen industry. The first large urea plants operated by the Government-controlled company, T. Pupuk Sriwidjaja (PUSRI) came on-stream in the 1970s and now the country has eleven nitrogen fertilizer plants, nine of these produce ammonia and urea and there are two plants producing ammonium sulphate. The total capacity of N is about 2.4 million tons. Several other plants are under construction or planned, which will bring capacity up to more than 3 million tons by the end of the decade. At Palembang in Sumatra, PUSRI has now three large and one small urea plants operating and there is an ASEAN a/ owned plant at Aceh. A further urea plant at Aceh operated by a state company, P. T. Iskander Muda has recently been commissioned. The Kujang Fertilizer Company operates a large urea complex at Citrampek and plans to bring another on stream at the end of the decade, while P. T. Pupuk Kalimantan (Kaltim) has recently brought two large plants on stream at Bontang and plans another for 1989.

---

a/ Association of South-East Asian Nations.

Indonesia

Nitrogen Fertilizer Production, Consumption and Trade ('000 tons N)

	1979/80	1980/81	1981/82	1982/83	1983/84	1984/85
Production	874	958	971	940	1,077	1,402
Consumption	620	851	997	1,082	1,049	1,285
Exports	138	83	18	44	148	104
Imports	14	125	184	235	134	128

1.107 Indonesia has large reserves of natural gas and the prospects for discovering more gas are excellent. The proven reserves are about 1,000 BCM of which about two-thirds is associated or dissolved gas. The first industrial use of gas came in 1964 with the commissioning of the PUSRI I fertilizer plant. The bulk of gas production presently comes from the huge non-associated gas reserves in Arun, North Sumatra and in Badak, East Kalimantan, which support the large Indonesian LNG projects. Indonesia is planning to increase its domestic utilization of natural gas in the fertilizer industry both for domestic and export uses.

Republic of Korea

1.108 In 1975/76, nitrogen fertilizer consumption in Korea reached 468,000 tons N. It fell to 361,000 the following year and although it climbed again to 462,000 in 1978/79, generally it has declined since. Part of this decline has been caused by readjustments in fertilizer N:P:K ratios in order to improve soil fertility. Because of the still high application rates of nearly 200 kg/Ha N, no significant increase in demand is expected in the future. There are seven nitrogenous fertilizer plants producing ammonia, ammonium nitrate, urea, and a range of N:P:K compounds. In 1980/81, South Korea had a capacity in excess of one million tons of nitrogen as ammonia, and exports of about 360,000 tons of

N. The ammonia plants in South Korea are based on naphtha feedstock and because of the high cost of this feedstock South Korea has become increasingly less competitive in the international market. The South Korean Government, through its Economic Planning Board, at the beginning of 1983 decided to make some capacity cuts and the plants at Chungju and Chin Hae were closed. This has reduced the nominal ammonia capacity to about 0.7 million tons. However, in a depressed international market one of the nitrogen plants at Yosu, although not permanently closed, has been idle during the last few years.

Republic of Korea

Nitrogen Fertilizer Production, Consumption and Trade ('000 tons N)

	1979/80	1980/81	1981/82	1982/83	1983/84	1984/85
Production	786	688	666	615	580	620
Consumption	444	447	432	309	367	402
Exports	279	362	185	193	210	234
Imports	-	-	2	1	2	4

1.109 All South Korean fertilizer plants currently use naphtha for ammonia production and are supplied by domestic petroleum refiners.

Malaysia

1.110 Fertilizer consumption has increased steadily in Malaysia over the last few years-from 78,000 tons of N in 1975/76 up to 166,000 tons of N in 1983/84. A Government investment body, PERNAS has recently acquired a number of large commercial plantations making the Government the largest fertilizer buyer in the country, accounting for as much as 65% of the country's total consumption. Malaysia's domestic production of ammonia originates from Exxon's small ammonia plant at Port Dickson which is sufficient to meet 90% of the country's ammonium sulphate and ammonium nitrate requirements. The Chemical Company of Malaysia (CCM) manufactures approximately 130,000 tons of ammonium nitrate product as an ingredient for N:P:K production. In 1983, FPM Sdn a joint venture between FELDA, KPM and Behn Meyer Sdn began production of granulated fertilizer based on urea as a nitrogen source. This plant has a capacity of 240,000 tons granular product per year. A large ammonia/urea complex which has recently come on stream in 1986 is the ASEAN Bintulu Fertilizer Plant, a joint venture of the five ASEAN Countries, Malaysia, Thailand, Indonesia, Philippines and Singapore. This plant has a daily production capacity of 1,000 tons of ammonia and 1,700 tons of urea. Malaysia will now become a large exporter of urea.

Malaysia

Nitrogen Fertilizer Production, Consumption and Trade ('000 tons N)

	1979/80	1980/81	1981/82	1982/83	1983/84	1984/85
Production	37	35	20	20	15	47
Consumption	138	139	128	138	235	249
Exports	-	-	-	-	-	-
Imports	140	110	105	142	190	194

1.111 The Exxon ammonia plant at Port Dickson is based on naphtha and refinery gas from the adjacent refinery complex. The ASEAN venture at Bintulu utilizes offshore gas from Shell's central Luconia field in the South China Sea. Malaysia's proven gas reserves are more than 1,400 BCM, of which about two-thirds are as non-associated. More than half of this is uncommitted and so there is good potential for building export-based ammonia plants in the future.

Philippines

1.112 Nitrogen fertilizer consumption increased steadily in the 1970s but fell in 1981/82 when subsidies were reduced. Demand, however, picked up in the following years. Following the closure of the Planters Products ammonia plant at Limay, Luzan, due to unfavorable feedstock costs, there is now no domestic production of nitrogen in the Philippines. The country is thus dependent entirely on imported ammonia and urea. In 1985, a large fertilizer factory owned by the Philippine Phosphate Fertilizer Corporation (PHILPHOS) came on stream. This plant will produce both diammonium phosphate and ammonium sulphate and will require increasing quantities of ammonia. Much of this will be imported from new plants in Indonesia.

Philippines

Nitrogen Fertilizer Production, Consumption and Trade ('000 tons N)

	1979/80	1980/81	1981/82	1982/83	1983/84	1984/85
Production	35	34	41	18	24	15
Consumption	227	225	211	231	240	178
Exports	-	-	-	-	-	-
Imports	217	224	113	222	190	178

1.113 There is no resource of natural gas in the Philippines and ammonia production has previously been based on naphtha or refinery gas.

Pakistan

1.114 During the 1960s and 1970s nitrogen fertilizer use grew rapidly, mainly due to a favorable subsidy program and support prices for crops. However, when the burden of subsidies became too great in 1982 and the Government started phasing them out, the growth rate began declining. It is expected, however, that once the situation has stabilized, demand should increase again although at a much lower rate than in the past, to reach about 1.1 million tons by 1990. There are four major nitrogen fertilizer producers in Pakistan; Dawood Hercules (Chichoki-Mallion); Esso (Daharki); Fauji (Goth Macchi); National Fertilizer Corp. (Multan, Hazira, Mirpur Mathelo). The country has six urea plants, one ammonium sulphate plant and a nitrophosphate plant producing compounds of 23-23-0 analysis. There are plans for a further ammonia plant to go on stream at Karachi at the end of the decade. With the large new Fauji plant coming on-stream in 1982 at a time when growth rate was declining, Pakistan has been a net exporter of nitrogen fertilizers in the last three years.

Pakistan

Nitrogen Fertilizer Production, Consumption and Trade ('000 tons N)

	1979/80	1980/81	1981/82	1982/83	1983/84	1984/85
Production	399	580	715	999	1,015	1,028
Consumption	776	844	933	953	914	935
Exports	-	-	-	45	158	210
Imports	44	34	95	131	75	87

1.115 Ammonia plants in Pakistan are based on domestic gas. There are 12 gas fields in Pakistan with reserves estimated at 510 BCM, most of which is non-associated. In view of existing ammonia commitments for this gas and other industrial needs, there may only be limited supplies of gas available for further ammonia capacity expansion.

Sri Lanka

1.116 Fertilizer consumption increased rapidly during the 1970s due to large subsidies but stagnated at the end of this period when subsidies were reduced. Demand picked up again when the new urea plant came on-stream in 1980/81. This plant was based on naphtha and unfortunately commissioning of the plant coincided with rapidly increasing naphtha prices and declining urea prices. The plant which has a capacity of 147,000 tpy N as urea has only operated intermittently since 1982 and in 1986 it was finally decided to close the plant.

Sri Lanka

Nitrogen Fertilizer Production, Consumption and Trade ('000 tons N)

	1979/80	1980/81	1981/82	1982/83	1983/84	1984/85
Production	-	-	38	91	58	64
Consumption	77	92	78	79	83	96
Exports	-	-	-	23	-	-
Imports	90	78	47	12	14	58

1.117 The ammonia/urea plant at Colombo, Sri Lanka was based on naphtha.

Thailand

1.118 Although nitrogen fertilizer consumption has increased steadily in Thailand, it is still a relatively small user of fertilizers. However, with the prospects of a new nitrogen fertilizer project and a promotional campaign to prepare the market for the increased domestic production, consumption has increased by 50% in the last two years. Thailand's only nitrogen fertilizer complex, with a capacity of 30,000 tpy ammonia based on lignite, closed in 1979. The country has relied on imports since then and will continue to import from the two ASEAN projects in Indonesia and Malaysia. The discovery of natural gas in the Gulf of Thailand has resulted in plans to use some of the gas for an ammonia/urea complex in the eastern seaboard area and the National Fertilizer Corporation has been established by the Government to implement the project, which was planned to come on-stream in 1989/90 but may now be delayed.

Thailand

Nitrogen Fertilizer Production, Consumption and Trade ('000 tons N)

	1979/80	1980/81	1981/82	1982/83	1983/84	1984/85
Production	-	-	-	-	-	-
Consumption	160	159	162	199	236	243
Exports	-	-	-	-	-	-
Imports	157	158	158	199	255	243

1.119 Reserves of non-associated gas in the Gulf of Thailand are in the central and southern sections of the Gulf in concession areas that belong to the Union Oil Company of Thailand and Texas Pacific Thailand. Total reserves are reported to be about 240 BCM. The proposed nitrogen fertilizer plant will be situated near the point where the main transmission pipeline makes its landfall. Because the cost of the gas is relatively high, it seems likely that it will be used mainly for ammonia and urea intended only for domestic consumption.

C. Centrally Planned Economies

1.120 Nitrogen fertilizer consumption and production has grown rapidly in both the Centrally Planned Economies of Eastern Europe and the Far East. The USSR and China are the two largest producers and consumers of nitrogen fertilizers. The USSR is the largest exporter of nitrogen and China the largest importer.

1. East Europe (including USSR)

1.121 This region is now the world's largest producer of ammonia and second largest user of nitrogen fertilizers. Overall the East Europe produces nearly 30% of the world's ammonia and consumes about 20% of the world's nitrogen fertilizers. It dominates the export market in both ammonia and urea. About two-thirds of the nitrogen industry in the USSR and the nitrogen industries in the other countries also use gas from the Soviet Union as feedstock.

Albania

1.122 This country consumes about 70,000 tons of nitrogen fertilizers as N in the form of urea and ammonium nitrate which is produced domestically at Fieri-Valona using fuel oil as a feedstock.

Bulgaria

1.123 Nitrogen fertilizer demand, spurred on by a relatively large domestic production, has increased steadily in recent years and now reached 0.5 million tons, equivalent to an application rate of more than 100 kg/ha. Nitrogen fertilizer production in 1984/85 exceeded 0.8 million tons, of which more than one third was exported. There are four major nitrogen fertilizer complexes in Bulgaria with large ammonia plants at Dimitrovgrad, Vratza, Varna and Stara Zagora. Fertilizer production includes both ammonium nitrate and urea with ammonium nitrate forming the major part of domestic consumption, and urea the major part of exports.

Bulgaria

Nitrogen Fertilizer Production, Consumption and Trade, ('000 tons N)

	1979/80	1980/81	1981/82	1982/83	1983/84	1984/85
Production	667	730	753	758	813	836
Consumption	421	450	518	536	550	479
Exports	216	247	200	190	232	301
Imports	6	6	6	5	6	9

1.124 Ammonia plants in Bulgaria are based on gas from the USSR through the Orenburg pipeline.

Czechoslovakia

1.125 Czechoslovakia has a well developed agricultural economy with one of the highest rates of fertilizer use in Eastern Europe. Demand for fertilizer nitrogen has been increasing steadily over the last decade or so at about 4% py, but is expected to slowdown through the remainder of the decade to about 2%. The country can almost meet its domestic needs from production at its two main production sites at Sala nad Vahom and Zaluzi, with ammonia capacity exceeding

0.8 million tpy N. A considerable amount of nitrogen fertilizer is imported but at the same time about 15% of domestic production is exported, mainly as urea. As no further capacity expansions are likely in the near future, exports are expected to decrease.

Czechoslovakia

Nitrogen Fertilizer Production, Consumption and Trade ('000 tons N)

	1979/80	1980/81	1981/82	1982/83	1983/84	1984/85
Production	650	705	674	656	686	713
Consumption	630	675	615	646	682	692
Exports	100	129	92	91	96	130
Imports	59	104	122	116	133	175

1.126 Although Czechoslovakia has some modest reserves of natural gas, like other Eastern European countries its nitrogen industry is based on natural gas from the USSR via the Orenburg pipeline.

German Democratic Republic

1.127 Nitrogen fertilizer consumption in the German D.R. appears to have developed and matured more quickly than most other East European countries and has been relatively stagnant in the last few years. There is also a significant demand of about 20% for the industrial sector mainly for ammonium nitrate for explosives and polymer intermediates. The main production units are at Schwedt, Leuna and Piesteriz. Capacity at the last site exceeds 740,000 tpy N in the form of two 1,360 tpd ammonia units and three 1,000 tpd urea plants. When this capacity was installed in the mid-1970s, East Germany became, and still remains a major exporter of nitrogen fertilizers, mainly as urea. Although an additional large new plant is being constructed at Rostock, this will be based on imported ammonia from the USSR and will be used to produce mainly calcium ammonium nitrate.

German Democratic Republic

Nitrogen Fertilizer Production, Consumption and Trade ('000 tons N)

	1979/80	1980/81	1981/82	1982/83	1983/84	1984/85
1983/842						
Production	874	943	967	948	968	959
Consumption	748	752	750	607	694	697
Exports	118	146	183	221	326	220
Imports	13	25	15	69	110	6

1.128 The East Germany nitrogen industry was originally based on lignite but was later converted to natural gas. Although the country produces significant quantities of natural gas, production of nitrogen in the German D.R. is based mainly on gas from the USSR.

Hungary

1.129 The Hungarian fertilizer market has reached maturity and there has been little growth in recent years. The production of nitrogen fertilizers more or less balances use, although there is also a considerable import and export trade in both ammonia and nitrogen fertilizers. There has been no new capacity since 1979 although a large new nitrogen plant is planned for Petfuerdoe, the large existing complex, within the next five year plan.

Hungary

Nitrogen Fertilizer Production, Consumption and Trade ('000 tons N)

	1979/80	1980/81	1981/82	1982/83	1983/84	1984/85
Production	640	651	691	671	700	688
Consumption	570	537	563	647	625	626
Exports	98	166	230	268	241	285
Imports	112	96	160	217	257	204

1.130 Ammonia production in Hungary is based on natural gas both from domestic resources and the USSR.

Romania

1.131 As in most other Eastern European countries, nitrogen fertilizer consumption increased rapidly between the mid-1960s and 1970s but the market has now reached maturity and only a small growth is expected in the future. The production of nitrogen fertilizers has also expanded considerably in Romania particularly during the period 1975-80 when several large fertilizer plants were constructed to exploit domestic natural gas. Romania is now the largest ammonia producer in the region after the USSR and has a capacity of 2.6 million tons of N as ammonia. As a result of this, the country is now one of the world's largest exporters of nitrogen fertilizers mainly in the form of urea and N:P:K compounds. The major nitrogen fertilizer complexes are at Tirgu Mures, Craiova, and Roznov. Romania now has to import natural gas from the USSR to meet its total needs and it seems unlikely that it will expand its nitrogen fertilizer production facilities further.

Romania

Nitrogen Fertilizer Production, Consumption and Trade ('000 tons N)

	1979/80	1980/81	1981/82	1982/83	1983/84	1984/85
Production	1,738	1,707	1,822	2,008	2,091	2,212
Consumption	786	646	882	884	878	857
Exports	838	1,015	1,125	1,037	1,300	1,303
Imports	-	-	-	-	-	-

1.132 Romania's nitrogen industry is all based on natural gas. Romania is the largest producer of natural gas in East Europe after the USSR but it also imports gas from the USSR.

Poland

1.133 Fertilizer demand in Poland grew rapidly during the 1970s but in recent years has declined slightly during the economic crisis facing the country. Nevertheless, it remains the second largest consumer of nitrogen fertilizer in the region after the USSR. Polish production almost matches demand, although there is also considerable trade in both exports and imports. The four major production complexes are at Tarnew, Kedzierzyn, Pulawy and Wlocklawek, where there is a yearly capacity of nearly 2 million tons N as ammonia. In recent years the operational efficiency of this capacity has been relatively low and there have been problems with the new fertilizer complex being constructed at Police which is now several years delayed.

Poland

Nitrogen Fertilizer Production, Consumption and Trade ('000 tons N)

	1979/80	1980/81	1981/82	1982/83	1983/84	1984/85
Production	1,376	1,290	1,274	1,298	1,342	1,369
Consumption	1,312	1,344	1,213	1,245	1,322	1,239
Exports	131	99	6	8	25	93
Imports	94	56	72	57	61	10

1.134 The ammonia plants in Poland are based on natural gas from domestic sources and the USSR.

USSR

1.135 The average application rates for nitrogen fertilizer in the USSR in 1984 was 44 kg per hectare of arable land and permanent crops. Although this figure looks low compared with Western Europe (109 kg/ha), it is of the same order of application as in the continent of North America which in 1984 was

53 kg/ha. The Soviet Union has tried hard to free itself from the need to import food by planning large increases in fertilizer production and application rates. Between 1955 and 1975, large increases were achieved in nitrogen fertilizer application but the rate of increase has fallen off considerably in later years mainly due to problems in construction delays in production facilities and also to major problems in the storage and distribution systems. The failings of the infrastructural system have been acknowledged by the CPSU Central Committee which in 1981 issued a resolution to strengthen the distribution system.

USSR - Nitrogen Fertilizer Consumption ('000 tons N)

1955/56	1960/61	1985/86	1970/71	1975/76	1980/81	1984/85
479	769	2,282	4,605	7,339	8,262	10,277

The 12th Plan period 1986-1990 calls for a total fertilizer use rate of about 30 million tons, a 50% increase in the rate achieved in 1982. This means an average increase of at least 5% per year through the remainder of the decade and if achieved could have a major impact on Soviet nitrogen products available for export.

1.136 The USSR is now the world's largest producer and exporter of ammonia and nitrogen fertilizers. With a large raw material resource base, the Soviet nitrogen fertilizer industry has expanded continuously over the last decade. During the 1970s the USSR ordered a total of 40 large scale ammonia plants with main contractors, Toyo Engineering of Japan, Chemico of the USA and Creusot-Loire of France. Most of these plants were due to be completed in the early 1980s but work has not proceeded as scheduled and some of the plants will not be completely commissioned until after the mid-1980s. Construction of urea plants in the USSR is proceeding in line with ammonia capacity but with some lag. Several large urea plants were scheduled to come on-stream in the mid-1980s and as they do, this impacts on the USSR export situation. With more nitrogen fertilizer capacity to meet its own growing needs the USSR may well have less ammonia to export.

1.137 It is difficult to assess the future potential of the Soviet nitrogen industry. Detailed plant lists are maintained by various analysts but these often vary considerably as plants' locations are specified under different names. Although the history of the new plants based on Western technology is fairly well known, information on the industry prior to the mid-1970s is not so well understood. By reference to production data for recent years published in the Comecon handbook it would appear that present ammonia plant utilization in the USSR is about 75% and so there is considerable margin to improve the output from existing plants in the future, in addition to any new plants which may be expected to come onstream.

USSR

Nitrogen Fertilizer Production, Consumption and Trade ('000 tons N)

	1979/80	1980/81	1981/82	1982/83	1983/84	1984/85
Production	9,074	10,155	10,581	11,481	12,687	13,143
Consumption	7,467	8,262	8,383	9,038	10,302	10,277
Exports	813	1,051	1,285	1,643	1,676	1,878
Imports	15	42	33	44	46	28

USSR - Ammonia Production and Exports ('000 tons N)

	1980	1981	1982	1983	1984
Production	13,840	14,760	14,650	17,655	18,600 a/
Exports	1,936	1,707	1,297	1,783	2,351

a/ Estimated.

1.138 The Soviet ammonia export commitments grew initially as a result of barter and buy-back deals with Western Europe and the US. The facilities have been constructed for exporting ammonia at Ventspils in the Baltic and at Odessa in the Black Sea. The main facility at Odessa is supplied from ammonia complexes at Toglyatti and Gorlovka by pipeline.

1.139 Up to 1960, about 80% of Russian ammonia capacity was based on coal or coke oven gas. In 1980 more than 90% was based on natural gas. The USSR with 36,000 BCM has the world's largest resources of natural gas and has built up a major industry for the export of fertilizers and petrochemicals based on gas. About 75% of Russian explored gas reserves are located in West Siberia and a major pipeline is now being constructed to bring this gas to the west. Other gas developments are taking place in Turkmenia and Astrakhan. The USSR has the potential to increase significantly its ammonia production based on natural gas and in view of its hard-currency earning capacity, there may be some incentive to expand its activities. The constraints will be, as in the past, its ability to construct and bring plants on-stream in a timely manner and the very low export prices in recent years will also serve as a deterrent.

2. Socialist Asia

1.140 This area comprises China, Kampuchea, Korea (DPR), Mongolia and Vietnam. China is by far the largest producer and consumer in the region whereas Mongolia and Kampuchea have no production facilities and only a small consumption.

China

1.141 China accounts for about 8% of the world's arable land but supports about 22% of the world's population. Since 1949 China has made tremendous progress in improving agricultural production, although sown area has remained relatively constant. During this period unit yields and food production have increased three to four times. Undoubtedly, one of the major contributors to this increase is the use of nitrogen fertilizer which has grown from virtually zero in 1949 to an application rate of about 130 kg/ha in 1983 equivalent to a total use of 11.5 million tons of N. In 1984/85 China had the world's largest usage of nitrogen fertilizer. Nitrogen fertilizer consumption in the future is expected to grow at a much reduced rate. Unless the use of other nutrients such as phosphate and potash can be increased significantly, large increases in nitrogen use will not reap full benefits.

1.142 Before 1949 two small plants produced an aggregate of 5,700 tons N as ammonium sulphate. Today there are about 1,400 nitrogen plants. The development of China's nitrogen fertilizer industry began mainly in the 1950s to produce ammonia almost exclusively from locally produced anthracite. There are now over 1,300 of these small plants, each producing 4,000-15,000 tons N py. The product is usually ammonia liquor and ammonium bicarbonate although a few of these plants also produce urea, ammonium chloride and ammonium nitrate. In 1983 the nationwide output of ammonia amounted to 13.92 million tons N of which large-scale plants produced 3.01 million tons accounting for 21.6% of the total, medium-scale plants 3.06 million tons, making up 22% and small-scale plants of 7.85 million tons equivalent to 56.4%. The main developments in plant capacity and output occurred in the 1970s when China purchased thirteen 1,000 tpd ammonia plants from the west. In addition about 50 medium-sized ammonia plants have been designed and built using Chinese resources. The products of the medium-sized plants are usually urea or ammonium nitrate. Even though ammonia production and urea has increased enormously in recent years, it has still not been sufficient to keep up with consumption and China has imported an increasing quantity of nitrogen fertilizer. In 1983 nitrogen fertilizers consumption in China comprised 30.3% urea, 58.0% ammonium bicarbonate 5.9% ammonium nitrate and 5.8% from other sources.

China

Nitrogen Fertilizer Production, Consumption and Trade a/ ('000 tons N)

	1979	1980	1981	1982	1983	1984	1985 b/
Production	8,820	9,993	9,858	10,219	11,094	12,211	11,584
Consumption	9,087	10,284	10,562	10,430	11,927	12,575	12,648
Exports	-	-	-	-	-	-	-
Imports	1,451	1,537	1,541	1,808	2,365	2,900	2,054

a/ Source: Ministry of Chemical Industry - China
b/ Provisional figures.

Note: These differ somewhat from FAO statistics where consumption is apparent consumption based on production plus imports.

1.143 Ammonia production is based on a wide range of feedstocks. In 1983 the total output of about 13.92 million tons of N as ammonia was made up as follows:

Solid feedstock (anthracite coal)	65.4%
Liquid feedstock (heavy fuel oil, naphtha)	13.8%
Gaseous feedstock (natural gas associated gas, coke oven gas refinery gas)	20.6%
Others	0.2%

For future production of ammonia, China has abundant resources of coal, oil and natural gas and, depending on location, China will continue to use all three main sources. Consideration is being given to using some of the potential offshore supply of natural gas at Hainan Island for nitrogen fertilizer production but by the time this project is completed the product could be destined for the export market.

Korea (DPR)

1.144 Consumption of nitrogen fertilizers in North Korea increased rapidly in the last part of the 1970s as new nitrogen fertilizer supply became available. In the last five years, however, consumption has not increased significantly. North Korea has about 870,000 tons of N capacity as ammonia and the two largest plants were added in 1976 and 1980 at Pyongyang and Changchangang, respectively. Most of the production is in the form of urea, although there are small plants producing ammonium sulphate and ammonium nitrate.

Korea (DPR)

Nitrogen Fertilizer Production, Consumption and Trade ('000 tons N)

	1979/80	1980/81	1981/82	1982/83	1983/84	1984/85
Production	550	553	600	588	608	620
Consumption	540	550	564	464	583	597
Exports	13	12	45	29	39	33
Imports	5	9	9	5	14	10

1.145 The feedstock for the plants is believed to be mainly naphtha.

Viet Nam

1.146 Due to the conflict and political changes in the area, fertilizer consumption has fluctuated considerably, although the last few years have seen asteady increase which is expected to continue through the 1980s. There is only one small production unit of about 50,000 tons of N as urea. A plant of 165,000 tpy N capacity is reported to be under construction and will start-up in the late-1980s.

Viet Nam

Nitrogen Fertilizer Production, Consumption and Trade ('000 tons N)

	1979/80	1980/81	1981/82	1982/83	1983/84	1984/85
Production	40	40	40	15	15	15
Consumption a/	121	156	200 a/	235 a/	351 a/	310 a/
Exports	-	-	-	-	-	-
Imports	79	114	160	220	274	295

a/ Apparent consumption based on production plus imports

1.147 No information is available on existing capacity. The planned capacity for 1986 is believed to be based on coal.

ANNEX 21. AMMONIA PRICES 1971-1986 (US Gulf)
(US$/ton f.o.b.)

Month	1971	1972	1973	1974	1975	1976	1977	1978	1979	1980	1981	1982	1983	1984	1985	1986
January	30-33	30-34	34-38	130-200	390-400	90-110	95-105	80-88	100-105	150-155	155-160	145-150	125-128	180-190	160-170	115-121
February	30-33	30-34	36-40	160-230	320-370	82-90	105-110	85-90	105-110	170-180	185-190	165-175	135-143	215-220	165-170	108-110
March	30-32	28-32	40-45	170-250	310-350	95-105	100-105	97-101	112-117	165-170	187-193	160-165	145-155	200-210	163-172	105-108
April	32-34	28-32	43-48	200-250	270-320	95-100	130-135	90-95	120-130	145-160	185-190	145-150	154-163	176-182	163-172	98-105
May	33-35	28-33	45-55	270-310	240-280	105-115	110-115	85-88	110-115	145-150	185-190	142-145	153-163	173-180	160-162	98-105
June	33-35	28-33	52-60	300-350	185-250	100-105	100-110	85-90	118-120	142-150	190-195	138-143	140-145	152-165	155-160	90-100
July	33-35	28-32	55-60	350-400	150-220	100-105	105-110	85-88	125-135	138-142	180-181	136-140	127-140	160-170	150-155	85-88
August	33-35	28-32	60-80	350-400	140-150	105-110	102-110	75-80	135-145	132-135	170-175	130-135	127-135	174-176	148-150	80-86
September	n.a.	28-32	65-80	350-400	130-150	100-105	102-110	85-90	138-143	132-137	160-162	126-132	137-142	175-180	148-150	80-83
October	31-33	29-32	70-90	350-420	100-120	100-110	93-98	85-90	140-145	132-137	155-158	126-130	150-160	175-180	140-146	76-79
November	31-34	30-33	75-120	380-420	105-120	102-105	88-95	80-85	136-138	132-136	145-150	126-128	160-175	168-178	123-126	77-80
December	30-33	32-36	100-170	370-400	90-110	90-100	83-90	80-85	135-140	145-150	140-145	123-128	160-164	163-172	122-125	75-76

Source: Industry Department
January 1987

ANNEX 21. UREA PRICES 1971-1986 (Bagged West Europe) (US$/ton f.o.b.)

Month	1971	1972	1973	1974	1975	1976	1977	1978	1979	1980	1981	1982	1983	1984	1985	1986
January	42-48	46-54	n/a	220-260	330-360	110-125	110-120	135-140	149-155	200-210	235-240	165-175	140-145	140-145	180-185	102-112
February	42-48	45-50	75-78	200-300	300-340	105-115	110-120	135-140	155-165	200-210	236-241	160-175	140-145	150-160	175-180	102-112
March	45-48	48-50	80-85	240-300	310-340	112-120	110-125	135-140	160-165	205-210	235-240	155-165	140-145	155-165	162-170	110-114
April	45-48	50-54	85-90	300-320	220-270	105-115	120-135	140-145	160-165	205-210	235-240	155-165	135-140	158-163	162-170	100-112
May	45-48	50-54	88-92	320-340	190-230	110-120	115-125	140-145	165-170	205-210	235-240	165-175	124-135	160-165	145-155	108-110
June	44-48	54-57	85-90	330-365	180-200	105-120	125-135	140-150	165-170	210-220	208-225	160-170	130-140	175-185	140-145	108-110
July	44-48	57-62	90-98	340-370	120-140	105-120	125-140	140-150	165-175	220-225	205-220	150-160	124-135	180-185	135-140	108-110
August	44-46	60-63	95-100	340-390	125-140	100-110	135-150	145-150	170-180	270-275	190-200	145-150	124-135	180-185	130-135	105-110
September	44-46	62-65	95-105	350-390	125-140	105-120	135-145	145-150	175-190	225-230	185-190	150-155	124-135	180-185	125-130	100-105
October	46-48	64-68	100-115	315-350	100-110	105-120	125-130	150-155	190-195	230-235	174-180	140-150	130-138	180-185	120-125	100-105
November	46-48	n/a	130-190	315-350	110-125	105-115	127-132	145-155	185-195	230-240	174-180	140-150	124-130	180-185	102-115	100-105
December	46-50	n/a	170-230	310-360	110-125	110-120	135-140	150-155	185-200	235-240	172-180	140-150	135-145	180-185	102-112	100-105

Source: Industry Department
January 1987

Distributors of World Bank Publications

Algeria: Office des Publications Universitaires, 1, place centrale de Ben-Aknoun, Algiers

Argentina: Carlos Hirsch, SRL, Galeria Guemes, Florida 165, 4th Floor-Ofc. 453/465, 1333 Buenos Aires

Australia, Papua New Guinea, Fiji, Solomon Islands, Vanuatu: Info-Line, Overseas Document Delivery, Box 506, GPO, Sydney, NSW 2001

Austria: Gerold and Co., A-1011 Wien, Graben 31, Austria

Bahrain: Middle East Marketing Research Bureau, Branch Office: Bahrain Research and Consultancy Associates W.L.L., P.O. Box 2750, Manama Town 317

Bangladesh: Micro Industries Development Assistance Society (MIDAS), G.P.O. Box 800, Dhaka

Belgium: Publications des Nations Unies, Av. du Roi 202, 1060 Brussels

Brazil: Publicacoes Tecnicas Internacionais Ltda., Rua Peixoto Gomide, 209, 01409 Sao Paulo, SP

Burma. See **Singapore**

Canada: Le Diffuseur, C.P. 85, 1501 rue Ampere, Boucherville, Quebec J4B 5E6

Chile: Editorial Renacimiento, Miraflores 354, Santiago

Colombia: Enlace Ltda., Aptdo. Aereo 34270, Bogota D.E.

Costa Rica: Libreria Trejos, Calle 11-13, Av. Fernandez Guell, San Jose

Cyprus: MEMRB Information Services, Head Office, P.O. Box 2098, Nicosia

Denmark: SamfundsLitteratur, Rosenoerns Alle 11, DK-1970 Frederiksberg C.

Dominican Republic: Editora Taller, C. por A., Isabel la Catolica a Esq. Restauracion, Apdo. postal 2190, Santo Domingo

Egypt, Arab Republic of: Al Ahram, Galaa Street, Cairo

Fiji. See **Australia**

Finland: Akateeminen Kirjakauppa, P.O. Box 128, SF-00101, Helsinki 10

France: World Bank Publications, 66, avenue d'Iéna, 75116 Paris

Germany, Federal Republic of: UNO-Verlag, D-5300 Bonn 1, Simrockstrasse 23

Greece: MEMRB Information Services, Branch Office, 24, Ippodamou Street, Athens-11635

Hong Kong, Macao: Asia 2000 Ltd., 6 Fl., 146 Prince Edward Road, W, Kowloon

Hungary: Kultura, P.O. Box 149, 1389 Budapest 62

India: New Delhi: *(for single titles)* UBS Publishers' Distributors Ltd., Post Box 7015, 5 Ansari Road, New Delhi 110002; *(for subscription series)* Universal Subscription Agency Pvt. Ltd. (Pending), 18-19 Community Centre Saket, New Delhi 110 017

UBS Branch Offices: 10 First Main Road, Gandhi Nagar, Bangalore 560009; Apeejay Chambers, P.O. Box 736, 5 Wallace Street, Bombay 400001; 8/1-B, Chowringhee Lane, Calcutta 700016; 7/188, 1(CA), Swarup Nagar, Kampu 208002; Sivaganga Road, Nungambakkam, Madras 600034

Indonesia: Pt. Indira Limited, Jl. Sam Ratulangi 37, Jakarta Pusat, PO Box 181

Ireland: TDC Publishers, 12 North Frederick Street, Dublin 1

Israel: Jerusalem Post, Jerusalem Post Building, P.O. Box 81, Romema Jerusalem 91000

Italy: Licosa, Libreria Commissionaria, Sansoni S.P.A., Via Lamarmora 45, Casella Postale 552, 50121 Florence

Ivory Coast: Centre d'Edition et de Diffusion Africaines (CEDA), 04 B.P. 541, Abidjan 04 Plateau

Japan: Eastern Book Service (EBS), 37-3, Hongo 3-Chome, Bunkyo-ku 113, Tokyo

Jordan, MEMRB Information Services, Branch Office, P.O. Box 3143, Jabal, Amman

Kenya: Africa Book Service (E.A.) Ltd., P.O. Box 45245, Nairobi

Korea, Republic of: Pan Korea Book Corporation, 134, 1-Ka, Shinmun-RO, Jongro-Ku, P.O. Box 101, Kwangwhamun, Seoul

Kuwait, MEMRB Information Services, Branch Office, P.O. Box 5465, Kuwait

Macao. See **Hong Kong**

Malaysia: University of Malaya, Cooperative Bookshop, Limited, P.O. Box 1127, Jalan Pantai Baru, Kuala Lumpur 59200

Maldives. See Sri Lanka

Mexico: INFOTEC, San Fernando No. 37, Col. Toriello Guerre, Thalpan

Morocco: Societe d'Etudes Marketing Marocaine, Branch Office, 2, rue Moliere, Casablanca

Netherlands: Medical Books Europe, BV (MBE), Noorderwal 38, 7241 BL Lochem

New Zealand: R. Hill and Son Ltd., Private Bag, New Market, Auckland

Nigeria: University Press Limited, Three Crowns Building Jericho, Private Mail Bag 5095, Ibadan

Northern Ireland. See **United Kingdom**

Norway: *For single titles:* Tanum-Karl Johan, A.S., Boks Box 1177 Sentrum, 0107, Oslo 1

Pakistan: Mirza Book Agency, 65, Shahrah-e-Quaid-e-Azam, P.O. Box No. 729, Lahore 3

Panama: Libreria Cultural Panamena, S.A. (LCP), Via Espana 16, Panama Zone 1

Papua New Guinea. See **Australia**

Peru: Editorial Desarrollo SA, Apartado 3824, Ica 242, OE. 106, Lima 1

Philippines: National Book Store, P.O. Box 1934, Manila

Portugal: Livraria Portugal, Rua Do Carmo 70-74, 1200 Lisbon

Saudi Arabia: Jarir Book Store, P.O. Box 3196, Riyadh—11471

Singapore, Taiwan, Burma: Information Publications, Private, Ltd., 02-06 First Floor, Pei-Fu Industrial Bldg., 24 New Industrial Road, Singapore 1953

Solomon Islands. See **Australia**

South Africa: *For single titles:* Oxford University Press, Southern Africa, Academic Division, P.O. Box 1141, Cape Town 8000. *For subscription series:* International Subscription Serv., P.O. Box 41095, Craighall, Johannesburg

Spain: Mundi-Prensa Libros, S.A., Castello 37, 28001 Madrid

Sri Lanka, Maldives: Lake House Bookshop, P.O. Box 244, 100, Sir Chittampalam A. Gardiner Mawatha, Colombo 2

Sweden: *For single titles:* ABCE Fritzes Kungl. Hovbokhandel, Regeringsgatan 12, Box 16356, S-103 27 Stockholm. *For subscription orders:* Wennergren-Williams AB, Box 30004, S-104 25 Stockholm

Switzerland: Librairie Payot, 6 Rue Grenus, Case postal 381, CH 1211 Geneva 11

Taiwan, China. See **Singapore**

Tanzania: Oxford University Press, P. O. Box 5299, Dar es Salaam

Thailand: Central Department Store, 306 Silom Road, Bangkok

Trinidad & Tobago: Systematics Studies Unit, 55 Eastern Main Road, Curepe, Trinidad

Tunisia: Societe Tunisienne de Diffusion, 5 Avenue de Carthage, Tunis

Turkey: Haset Kitabevi A.S., 469, Istiklal Caddesi, Beyoglu-Istanbul

Uganda: *Shipping:* Uganda Bookshop, P.O. Box 7145, Kampala

United Arab Emirates: Middle East Marketing Research Bureau, Branch Office, P. O. Box 6097, Sharjah

United Kingdom and Northern Ireland: Microinfo Ltd., P.O. Box 3, Alton, Hampshire GU 34 2PG

Vanuatu. See **Australia**

Venezuela: Libreria del Este, Aptdo. 60.337, Caracas 1060-A

Western Samoa: Wesley Bookshop, P.O. Box 207, Apia

Yugoslavia: Jugoslovenska Knjiga, YU-11000 BelgradeTrg Republike, 8th Floor

Zimbabwe: Textbook Sales Pvt. Ltd., Box 3799, Harare